The Western Bark Beetle Research Group: A Unique Collaboration With Forest Health Protection

Proceedings of a Symposium at the 2007
Society of American Foresters Conference

October 23–28, 2007, Portland, Oregon

Compilers

Jane L. Hayes is a research biological scientist, USDA Forest Service, Pacific Northwest Research Station, Forestry and Range Sciences Laboratory, 1401 Gekeler Lane, La Grande, OR 97850. **John E. Lundquist** is an entomologist, USDA Forest Service, Pacific Northwest Research Station, Forestry Sciences Laboratory, 3301 C Street, Suite 200, Anchorage, AK 99503.

Papers were provided by the authors in camera-ready form for printing. Authors are responsible for the content and accuracy. Opinions expressed may not necessarily reflect the position of the U.S. Department of Agriculture.

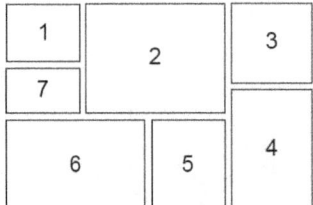

Cover photographs: (1) white spruce in Alaska, by William M. Ciesla, Forest Health Management International; (2) western pine beetle (*Dendroctonus brevicomis*), by Erich G. Vallery, USDA Forest Service; (3) residential area in California, by Chris Fettig, USDA Forest Service; (4) lodgepole pine in Colorado, by William M. Ciesla, Forest Health Management International; (5) residential area in California, by Laura Merrill, USDA Forest Service; (6) Douglas-fir in Idaho, by Ladd Livingston, Idaho Department of Lands; (7) mountain pine beetle larvae (*Dendroctonus ponderosae*), by Scott Tunnock, USDA Forest Service. Photos 1, 2, 4, 6, and 7 courtesy of Bugwood.org.

The Western Bark Beetle Research Group: A Unique Collaboration With Forest Health Protection

Proceedings of a Symposium at the 2007 Society of American Foresters Conference

October 23–28, 2007, Portland, Oregon

J.L. Hayes and J.E. Lundquist, compilers

U.S. Department of Agriculture, Forest Service
Pacific Northwest Research Station
Portland, Oregon
General Technical Report PNW-GTR-784
April 2009

Abstract

Hayes, J.L.; Lundquist, J.E., comps. 2009. The Western Bark Beetle Research Group: a unique collaboration with Forest Health Protection—proceedings of a symposium at the 2007 Society of American Foresters conference. Gen. Tech. Rep. PNW-GTR-784. Portland, OR: U.S. Department of Agriculture, Forest Service, Pacific Northwest Research Station. 134 p.

The compilation of papers in this proceedings is based on a symposium sponsored by the Insect and Diseases Working Group (D5) at the 2007 Society of American Foresters (SAF) convention in Portland, Oregon. The selection of topics parallels the research priorities of the Western Bark Beetle Research Group (WBBRG) (USDA Forest Service, Research and Development), which had been recently formed at the time of the symposium. Reflecting a unique partnership within the Forest Service, each paper was jointly prepared by a research scientist with the WBBRG and one or more entomologists with Forest Health Protection (USDA Forest Service, State and Private Forestry). Among these papers is a description of the currently elevated impacts of bark beetles in the Western United States; descriptions of the current state of knowledge of bark beetle response to vegetation management and also to climate change; discussions of the complex interactions of bark beetles and fire and of the complex ecological and socioeconomic impacts of infestations; an overview of the use of semiochemical (behavioral chemicals)-based technology for conifer protection; and a case study exemplifying efforts to assess risks posed by nonnative invasive bark beetles.

Keywords: Bark beetles, vegetation management, climate change, fire, socioeconomic impacts, semiochemicals, risk assessment.

Preface

Making complex decisions about insect pests involving multiple objectives and multiple criteria is not new to forest managers, but the need for systematic and scientific methods of decisionmaking has never been greater. Nothing illustrates this need better than the strikingly elevated levels of bark-beetle-caused tree mortality in forests of the Western United States during the last decade. The increasing challenges of addressing this issue in an environment of shrinking resources spawned the formation of the Western Bark Beetle Research Group (WBBRG), which comprises the research entomologists from the three Western USDA Forest Service R&D research stations.

The compilation of papers in this proceedings is based on a symposium at the 2007 Society of American Foresters (SAF) convention in Portland, Oregon. The selection of topics parallels the research priority list of the WBBRG, which had been recently formed at the time of the symposium. The aim of the symposium was to describe the currently elevated impacts of bark beetles in the Western United States and to showcase the significant efforts by the Forest Service to understand, manage, and mitigate these impacts through basic and applied research. The symposium was sponsored by the D5 Insect and Diseases Working Group of the Society of American Foresters. It is a long-term objective of both the WBBRG and the SAF D5 Working Group to enhance communication with their partners and stakeholders. This symposium represents one step taken by both groups to achieve this common goal.

The WBBRG serves to enrich interactions among bark beetle researchers and their partners. Cooperative research and the team approach are integrated into the concept of this group. As a consequence, work of the WBBRG involves a variety of partners, primarily the Forest Health Protection (FHP) staff of the USDA Forest Service State and Private Forestry branch. Accordingly, for the symposium, a research station scientist was teamed up with an FHP entomologist and asked to describe current research and how it relates to current management issues. Similarly, the collection of papers from the symposium in this proceedings cut across a range of the current most rapidly advancing topics in bark beetle research and the most urgent management issues in pest management in the Western United States. The proceedings papers also illustrate some of the emerging challenges faced by forest entomologists.

We gratefully acknowledge the Society of American Foresters for its assistance in planning and presenting the symposium. We also acknowledge the Pacific Northwest Research Station, the Pacific Southwest Research Station, the Rocky Mountain Research Station, and the Forest Health Protection staffs from Regions 1, 2, 3, 4, and 10 for their assistance. All of the papers were peer reviewed, and we are especially thankful for the numerous people that served as reviewers. Most of all, we appreciate the perceptive insights and state-of-the-art knowledge generously shared by the authors.

Jane L. Hayes and John E. Lundquist, Compilers

Contents

Bark Beetle Conditions in Western Forests and Formation of the Western Bark Beetle Research Group[1]

Robert J. Cain and Jane L. Hayes[2]

Abstract

The recent dramatic impacts of bark beetle outbreaks across conifer forests of the West have been mapped and reported by entomology and pathology professionals with Forest Health Protection (FHP), a component of USDA Forest Service's State and Private Forestry, and their state counterparts. These forest conditions set the stage for the formation of the Western Bark Beetle Research Group (WBBRG), comprised of research scientists within the three western research stations of the USDA Forest Service Research and Development. Facing the increasing bark beetles impacts, the newly formed WBBRG, in concert with FHP professionals from the western Regions, developed research priorities. Building on a strong foundation of past and present research, WBBRG scientists in conjunction with their varied partners will investigate the complex interactions of bark beetles and their hosts. Interactions to be explored include those within vegetation management scenarios at the individual tree to landscape scale, those between wildland fire and bark beetles, the long-term impacts of bark beetle outbreaks on ecological and socioeconomic values, and importantly the response of bark beetle systems (i.e., bark beetles, their hosts and common associates) to climate change. This increased understanding of bark beetle behavior and population dynamics at multiple scales and with other agents of change will lead to the development and improvement of management tools. As in the past, WBBRG scientists will work closely with FHP entomologists to implement practical research products to prevent, retard, or suppress unwanted effects of native and nonnative invasive bark and woodboring beetles in the West.

Keywords: Aerial survey, Forest Health Protection, Western Bark Beetle Research Group.

---

[1] The genesis of this manuscript was a presentation by the authors at the Western Bark Beetle Research Group—A Unique Collaboration with Forest Health Protection Symposium, Society of American Foresters Conference, 23-28 October 2007, Portland, OR.

[2] **Robert J. Cain** is an Entomologist, USDA Forest Service, R-2 Forest Health Protection, 740 Simms Street, Golden, CO 80401; email: rcain@fs.fed.us. **Jane L. Hayes** is a Research Biological Scientist, USDA Forest Service, Pacific Northwest Research Station, Forestry and Range Sciences Laboratory, 1401 Gekeler Lane, La Grande, OR 97850; email: jlhayes@fs.fed.us.

Introduction

As background for the presentations given at the 2007 SAF Conference Symposium, Western Bark Beetle Research Group—A Unique Collaboration with Forest Health Protection and the collection of papers in this Proceedings of that session, we describe the current trends in bark beetle-caused tree mortality in western forests. The many research challenges presented by these conditions provided compelling motivation for establishing a new west-wide Forest Service research group focusing on this situation. We outline the priority research topics defined by the group at their inaugural meeting with consensus by Regional partners. Past and present research experience and accomplishments that helped shape these priorities are briefly described.

Bark Beetle Conditions in Western Forests

Bark beetles have been causing dramatic tree mortality and making headlines across the West in recent years. Entomologists and pathologists with Forest Health Protection (FHP), a component of the USDA Forest Service State and Private Forestry (S&PF) and their state counterparts annually report insect and disease conditions. Acres affected by bark beetles across western forests are assessed through the creation of aerial survey sketchmaps. From fixed wing aircraft such as a Cessna 206, sketchmappers record polygons of insect activity in forest stands on USGS maps or on computer touch screens while the plane is flown along contours or predetermined flight lines. The tree species impacted, the damaging agent and the intensity are indicated for each polygon. When the damaging agent is a bark beetle, the intensity is determined by estimating the number of trees per acre that are currently fading. This becomes more difficult in large outbreaks with multiple years of damage and often multiple damaging agents active in the same area.

In recent years there have been widespread outbreaks of bark beetles across western North America. Outbreaks of native bark beetles have occurred across forest types from the low elevation pinyon-juniper woodlands to high elevation Engelmann spruce and subalpine fir forests (USDA 2005). Table 1 lists many of the bark beetles that have caused mortality over thousands of acres of their respective hosts. Native bark beetle populations are most influenced by stand conditions and weather conditions. Generally, older denser stands with larger trees and warmer, drier conditions are more favorable to bark beetles. Figure 1 shows the majority of the major forest cover types in the Rocky Mountain Region are over 100 years old and this is representative of conditions across the West.

Table 1—Western bark beetle species that have caused significant tree mortality in the last 10 years

Bark Beetle(s)	Host(s)
Spruce beetles, *Dendroctonus rufipennis* (Kirby)	Engelmann spruce (*Picea engelmannii* Parry ex Engelm.), white spruce (*P. glauca* [Moench] Voss), Sitka spruce (*P. sitchensis* [Bong.] Carr.)
Pinyon ips, *Ips confusus* (LeConte)	Pinyon pine (*Pinus edulis* Engelm. and *P. monophylla* Torr. & Frem.) and others
Pine engraver, *Ips pini* (Say), Arizona five spined ips, *Ips lecontei* Swaine	Ponderosa pine (*Pinus ponderosa* C. Lawson)
Western pine beetle, *Dendroctonus brevicomis* LeConte	Ponderosa pine, Coulter pine (*Pinus coulteri* D. Don)
Jeffrey pine beetle, *Dendroctonus jeffreyi* Hopkins	Jeffrey pine (*Pinus jeffreyi* Balf.)
Mountain pine beetle, *Dendroctonus ponderosae* Hopkins	Ponderosa pine, lodgepole pine (*P. contorta* Douglas ex Louden), white pines and others (*Pinus* spp.)
Douglas-fir beetle, *Dendroctonus pseudotsugae* Hopkins	Douglas-fir (*Pseudotsuga menziesii* (Mirb.) Franco)
Fir engraver beetle, *Scolytus ventralis* LeConte	True firs (*Abies* spp.)
Western balsam bark beetle, *Dryocoetes confusus*, Swaine	Subalpine fir (*Abies lasiocarpa* (Hook.) Nutt.)

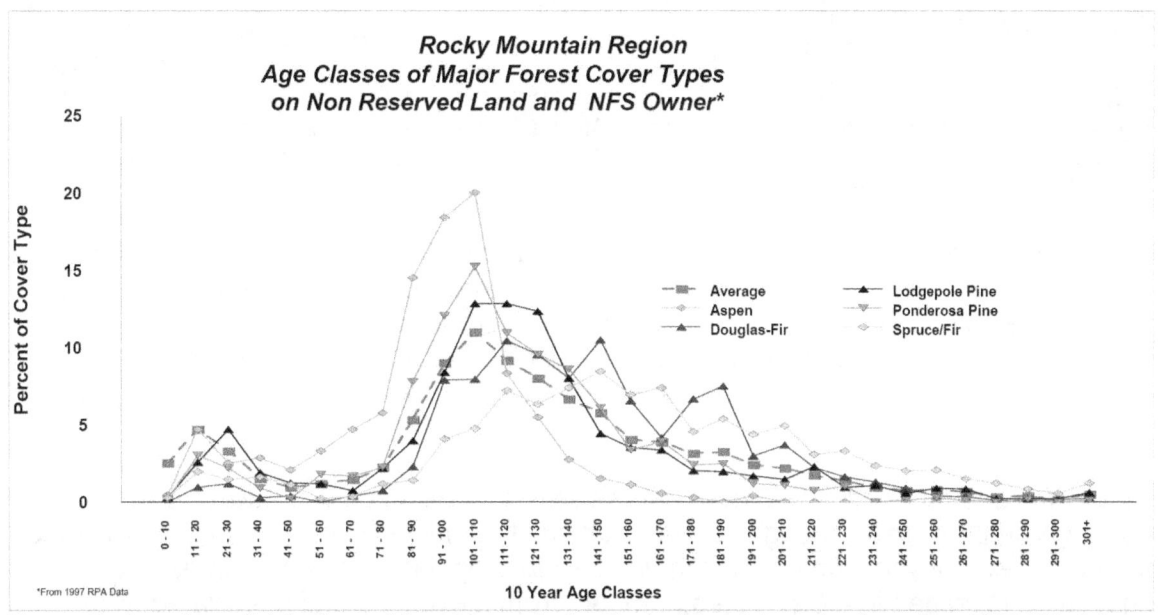

Figure 1—Age class distributions of forest types in the Rocky Mountain region based on 1990 FIA data.

Spruce beetle—Through the 1990s the largest spruce beetle epidemic ever recorded in North America eventually impacted to varying degrees over 3.2 million acres in Alaska including 1.4 million acres on the populated and extensively visited Kenai Peninsula (figure 2). This epidemic triggered some of the early widespread speculation in the media about the ecological impacts of warmer global temperatures (Juday 1998). Research has subsequently confirmed the connection between increased temperatures and spruce beetle population build-up (Hansen et al. 2001, Berg et al. 2006).

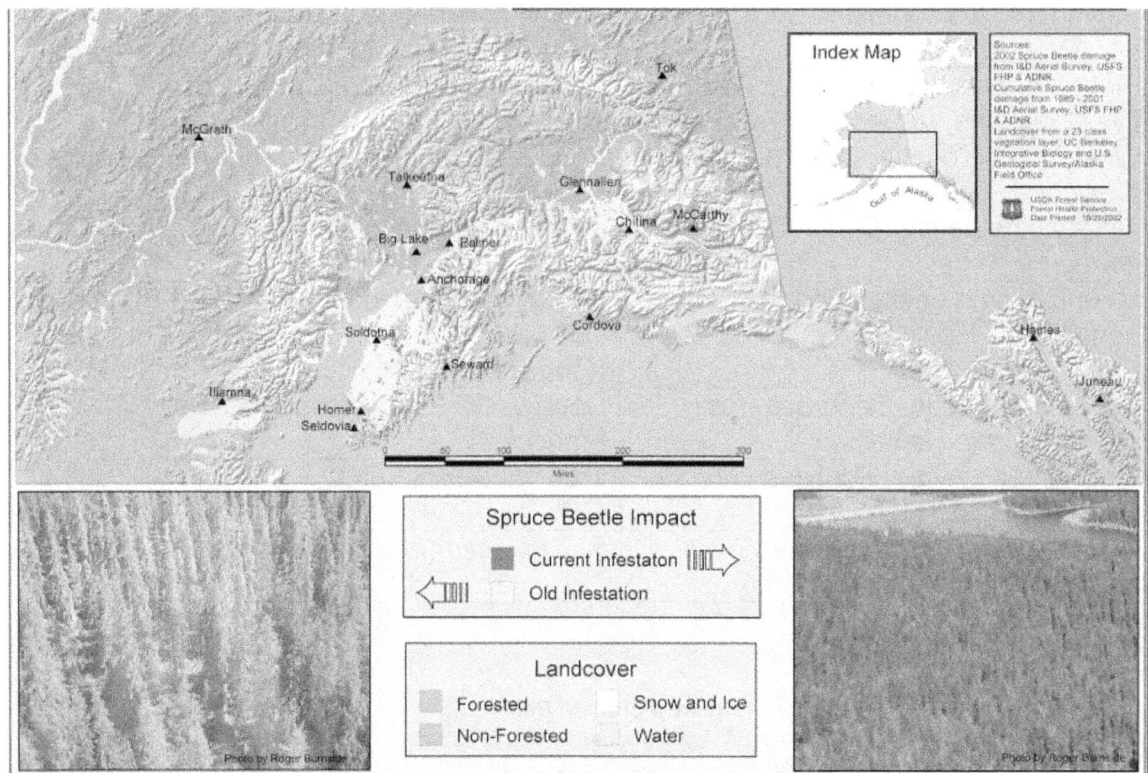

Figure 2—Spruce beetle outbreak in south-central Alaska 1989–2002 (1989–2001 in yellow, 2002 in red). Source: http://www.fs.fed.us/r10/spf/fhp/Condrpt03/2003%20Web%20Maps/slides/Spruce%20Beetle%20 Outbreak%20-%202003.html.

Spruce beetle has been active in other western states as well. Strong winds that blew down high elevation stands of Engelmann spruce created suitable host material that favored the build-up of spruce beetle populations. Outbreaks were first noted throughout Utah, then Colorado and Wyoming in the 1990's and 2000's. Much of Utah's spruce forests have been killed and areas of tree mortality continue to increase in Colorado and Wyoming.

Pinyon ips and other bark beetles in southwestern pines—The late 1990's and early 2000's brought extreme drought to the Southwest combined with warmer than average temperatures. Pinyon pines, although adapted to irregular moisture regimes and shallow soils, began to die in record numbers from pinyon ips and associated twig beetles (Breshears et al. 2005). Although scattered references exist to another large die off in the 1950s, there were many areas of large pinyons that had survived the 1950s drought that succumbed in the 2000s. The impact was felt over six states and over 650,000 acres were affected. Improved moisture conditions by 2004 helped to end the pinyon ips epidemic.

During that same drought period in the Southwest, large areas of ponderosa pine forests in central Arizona were killed by the Arizona five-spined ips and associated bark beetles. Also, southern California's Angeles, San Bernardino and Cleveland National Forests and adjacent land experienced extremely high levels of tree mortality due to a complex of native bark beetles, dense stand conditions and severe drought. During 2003–2004, western pine beetles, Jeffrey pine beetles and mountain pine beetles all contributed to the dying trees that appeared on the landscape in and around resort communities like Arrowhead Lake. In 2003, massive wildfires driven by Santa Ana winds burned through chaparral, homes, and forested areas in which bark beetle killed trees were prevalent. Moisture conditions throughout the southwestern United States improved in many areas and bark beetle activity decreased.

Mountain pine beetle—Mountain pine beetle is currently making the most dramatic widespread changes on the landscape across the West. These beetles were first described at the turn of the last century in the Black Hills of South Dakota. A large outbreak was occurring at that time and the following unattributed quote was found on an archived slide at the USFS's Forest Health Office in Lakewood, CO. "At the time of their flight, they settled on cabins like swarms of locusts". Today, just over 100 years later the ponderosa pine forests of the Black Hills are again experiencing an intensifying mountain pine outbreak that is making dramatic landscape changes.

In recent years, mountain pine beetle has impacted millions of acres of lodgepole pine forests across the West at levels not previously recorded. If you look at the range of lodgepole pine in North American and the cumulative map of acres impacted from 2002–2006 you can see that most of the lodgepole pine cover type has been impacted (figure 3).

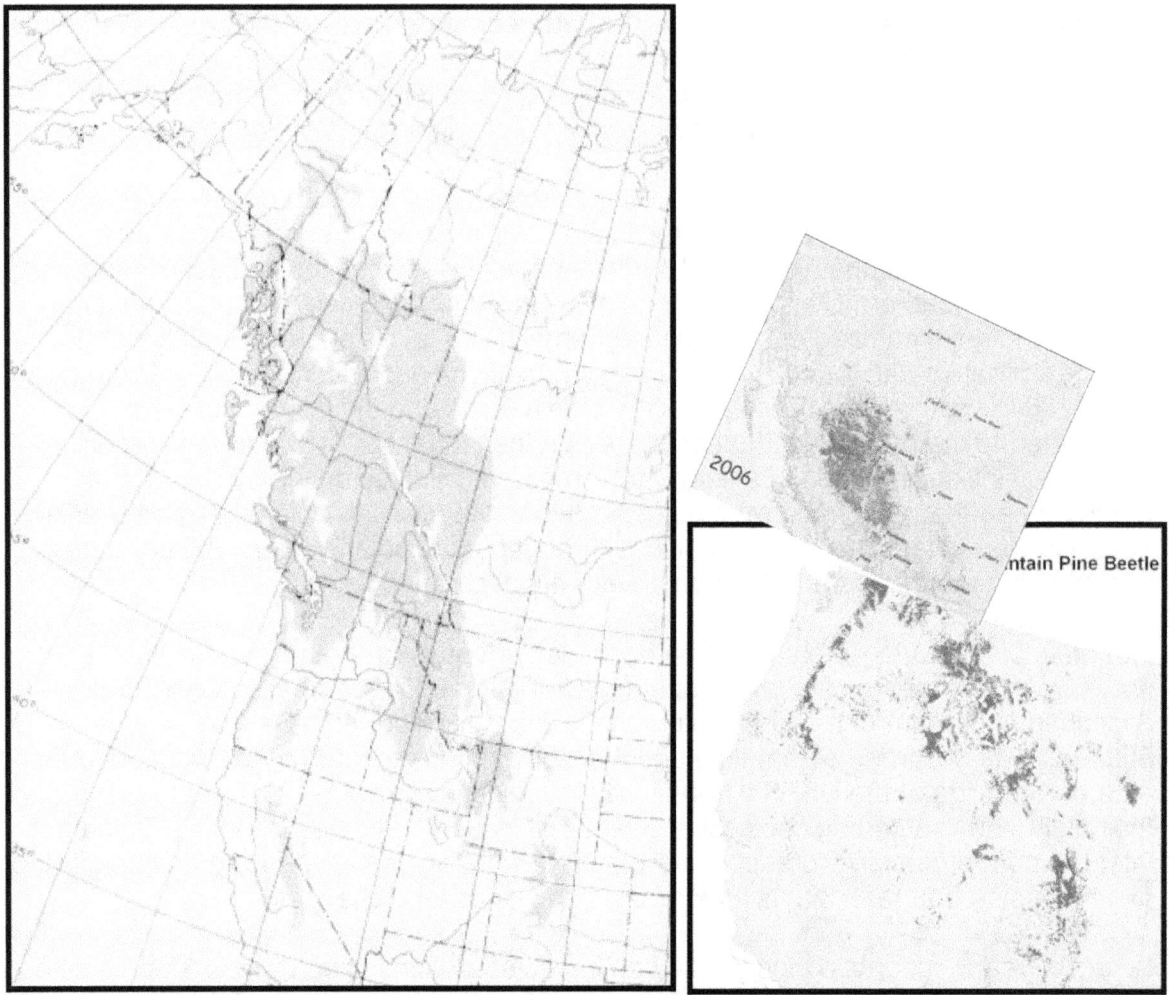

Figure 3—Range map of lodgepole pine in North America and mountain pine beetle impacted areas detected in aerial surveys. Canadian map is from the Canadian Forest Service website, http://mpb.cfs.nrcan.gc.ca/map_e.html, areas impacted by mountain pine beetle in 2006. The western U.S. map shows areas impacted by mountain pine beetle from 2004–2006.

The current epidemic has impacted stands at higher elevations and latitudes than have been previously recorded. In the Yellowstone corner of Montana, Wyoming and Idaho, high elevation white bark pines stands are being killed on sites previously considered to be too cold for serious mountain pine beetle epidemics (figure 4). Younger stands in old clearcuts, burns and avalanche runs remain green, but trees that are over five or six inches in diameter at the base are being killed by beetles attacking low on the trunk (figure 5). Twig beetles are also found attacking these smaller trees in regenerating stands in Colorado. Their populations may be building in the smaller diameter portions of trees attacked and killed by mountain pine beetle.

Figure 4—High elevation white bark pine killed in Idaho. Photo provided by Carl Jorgensen.

Figure 5—Mature lodgepole pine killed by mountain pine beetle in north central Colorado. Most of the trees regenerating in the old clearcut are too small for the beetles, however, many trees reaching five to six inches in diameter at the base are being killed by mountain pine beetles and twig beetles. Photo by Sheryl Costello.

The mountain pine beetle epidemic in northern Colorado illustrates how quickly beetle populations can increase and impact extensive areas. The first signs of a building beetle population occurred in 1997. By 1999, clearly defined epicenters were mapped during aerial surveys. These rapidly expanded and merged and by 2006 most of the cover type west of the Continental Divide had some level of mountain pine beetle activity. Cole and Amman (1980) reported that over the course of an outbreak, most of the large diameter trees will be killed by the time the outbreak subsides. They also reported that an outbreak averages six years to run its course in a given stand, but emphasized that once infestations build up, a large amount of dispersal may occur. This leads to more rapid tree losses in adjacent stands with beetle populations reaching outbreak levels and subsiding in a shorter time. Some newly infested areas are now being depleted of suitable host trees in only one or two years. Through 2006, the cumulative area of lodgepole pine forests in Colorado where mountain pine beetle activity was detected in aerial surveys was about 1 million acres. In 2007, that number increased to 1.5 million acres. The epidemic in British Columbia, where lodgepole forests are more contiguous than in the western U.S., is even more staggering. The B.C. Ministry of Forests and

Range reported over 24 million acres of lodgepole pine affected by mountain pine beetle in 2007 (Buxton unpublished).

Douglas-fir beetle—Visitors to Yellowstone National Park will notice old Douglas-fir trees killed by Douglas-fir beetle along the road from Cody, Wyoming to the east entrance of the park. Douglas-fir beetle outbreaks followed forest fires and drought and peaked in 2005 when over 670,000 acres were affected across the West. Acres of Douglas-fir tree mortality have been declining, but in 2007, increases in Douglas-fir beetle activity were recorded in the central and southern Rockies, intermountain west and the Pacific Northwest.

Western balsam bark beetle and fir engraver beetle—Western balsam bark beetle primarily attacks the high elevation subalpine firs and fir engraver beetles are most common on the other western true firs. Notable outbreaks have occurred in recent years across the West and are closely tied to local drought conditions. Outbreaks tend to subside when soil moisture improves.

There are also many areas in the West where more than one bark beetle species is active at the same time. On the Shoshone National Forest in northern Wyoming, uniquely pure stands of Rocky Mountain Douglas-fir are being killed by Douglas-fir beetle next to stands of Engelmann spruce killed by spruce beetle and limber and lodgepole pine being killed by mountain pine beetle. This scenario is being repeated throughout the West, in different types of host stands and with different beetle species.

Given these recent trends and present level of beetle-caused tree mortality, it is not difficult to see how the view out the window of a plane or even a car may lead one to the conclusion that western conifer forests are under attack. Certainly, vast areas of the western landscape have been affected by western bark beetle infestations—outbreaks involving several stands to epidemics encompassing a host type across multipe forests. Millions of acres are considered at risk (Western Forestry Leadership Coalition 2007). Whether the levels we see today are historically unprecedented is subject to debate. We lack sufficient records to adequately address the issue, although we know there were large-scale bark beetle infestations at the turn of the twentieth century when forest entomologists first began studying the insects of western forests (e.g., Wickman 2005) and other epidemics since then are well documented. Nevertheless, given the relationship of recent losses to changing climatic conditions, there exists a threat of increasing tree losses with projected climate changes. The current western bark beetle situation presents many opportunities for research to better understand changing western forest ecosystems and the management implications of this large disturbance, its source and interactions with other agents of change.

Formation of the Western Bark Beetle Research Group

These conditions were a driving force in bringing together research entomologists of the three western Forest Service research stations to form the USDA FS R&D Western Bark Beetle Research Group (WBBRG). The leadership of the Pacific Northwest, Pacific

Southwest and Rocky Mountain Research Stations, which cover the 15 western states including Hawaii and Alaska, recognized the bark beetle situation in the West as a compelling problem that crosses station boundaries. The WBBRG is made up of 11 researchers located from Alaska to Arizona, whose work focuses on native and non-native insects of western forest and rangeland ecosystems. With many research challenges, the benefits of a tri-Station partnership include improved efficiency by leveraging resources and expertise, and enhanced communication and coordination.

The WBBRG serves as an ad hoc umbrella organization aimed at fostering communication, enhancing responsiveness and delivery of bark beetle research, and enriching scientific interactions among Forest Service bark beetle researchers in the western U.S. The objectives of this group include:

- Work with partners and stakeholders to identify western bark beetle priority research
- Pursue priority research and develop high impact products
- Promote the relevance of western bark beetle research for partners and stakeholders
- Increase overall quality, productivity and timeliness of research through cooperation and integration among stations
- Enhance communication and service to partners and stakeholders

To achieve the first of these objectives, the WBBRG invited forest entomologists from FHP representing the western Regions to participate in this endeavor (see also Negrón et al. 2008b). When the ideas were synthesized, the consensus was that among the numerous research topics raised, the following represent the highest ranked priorities:

- Describe, evaluate, and quantify long-term outcomes of bark beetle outbreaks on ecological, economic, and social services at various spatial scales.
- Evaluate bark beetle response to vegetation treatments at the tree, stand, and landscape levels.
- Determine the relationships between bark beetles and wildfire.
- Evaluate bark beetle, common associates, and host tree physiological responses to climate change.
- Develop new and improved chemical and semiochemical-based strategies for bark beetle management.
- Develop methods and strategies for detecting, monitoring, and eradicating or mitigating invasive bark beetles and woodboring insects.

Part of a Long History of Western Forest Entomology Research
To accomplish these goals, the WBBRG is continuing to build on past research successes. Often teamed with FHP entomologists or other partners, FS R&D entomologists of the three western stations have a long history of conducting research that is relevant to land managers and owners. Forest insect research, especially bark beetle research, has had a prominent role in FS R&D in the West over the years and made significant contributions. Since the turn of the last century professional forest entomologists have been conducting research and sharing their findings and knowledge

with colleagues, partners, and clients. In 1899, A.D. Hopkins (commonly known as the father of forest entomology in the U.S.) made a 2-month trip to the Pacific Northwest. The "Preliminary Report of the Insect Enemies of Forests in the Northwest" from that trip arguably marks the beginning of forest entomological research in the West (Burke 1946 in Wickman 2005). Soon after this trip Hopkins, a bark beetle expert, became the first Chief of the Division of Forest Insect Investigation established in 1902. He made subsequent trips (1902–1905) to the bark beetle outbreak in the Black Hills, to Colorado, to the southwest, and other parts of the Pacific Slope, and eventually a visit in 1911 to the Northeastern Oregon Project, the first large-scale bark beetle control project in the West (Wickman et al. 2002).

It was in large part the dominant role of bark beetles in forests of the West and elsewhere that led to the creation of the Division of Forest Insect Investigation within the USDA Bureau of Entomology to work with the Bureau of Forestry, headed at the time by Gifford Pinchot. The Division established stations throughout the West including the Pacific Slope (eventually settling at UC Berkeley and Portland, OR), Fort Collins, CO, Coeur'd Alene, ID, and Missoula, MT. The Division pursued bark beetle and other entomological research until 1953 when it officially became a part of the USDA Forest Service and its functions were transferred to Forest and Range Experiment Stations, which eventually became Research Stations. Early activities of the Division naturally focused on identifying the insects of greatest concern, including studies of taxonomy and biology, and developing methods for control. Scientists with the Division played significant roles in cooperative bark beetle control projects with other agencies (e.g., Forest and Park Services) and private landowners in the West, such as the Northeastern Oregon Project (1910–11) and others into the 1930s, particularly in California and Oregon (Wickman et al. 2002). In later years, as more permanent laboratory facilities were established, the focus of bark beetle research shifted to ecological investigations and control of bark beetles through forest management practices.

Bark beetles and vegetation management—Studies by early researchers laid groundwork for the research of today. When it had become clear that direct control methods used in large-scale control projects were having little long-term impacts on reducing levels of bark beetle-caused tree mortality, they shifted their attention to silvicultural and forest management strategies. For example, a tree susceptibility classification system was developed in 1942 (Keen 1943), leading the way for considerable future research and development of stand hazard- or risk-rating systems that help managers identify stand susceptibility and the probability of bark beetle infestation (e.g., Schmid and Frye 1976, Stevens et al.1980). Many of these systems, or updated successors, are still widely used at the project level to guide silvicultural and restoration treatments and some research work has continued in the area where gaps exist (Negrón 1997, 1998, Negrón and Popp 2004, Negrón et al. 2008a). Response of beetles to vegetation treatments has been a subject of past research (reviewed by Fettig et al. 2007); however, particularly in light of the current emphases on fuel reduction and forest restoration, sufficient knowledge gaps exist to be ranked as an area of high priority research by WBBRG. For a more detailed description of this topic, see McMillin and Fettig in this Proceedings.

Long-term consequences of bark beetles impacts on ecological and socioeconomic values—Previously comprehensive reviews or annotated bibliographies of research on some of the most significant bark beetles were published (western pine beetle, Miller and Keen 1960; Douglas-fir beetle, Furniss 1979; mountain pine beetle, Lessard et al. 1986; spruce beetle, Linton and Safranyik 1987). As forest management has shifted to multiple resource management, bark beetle research has also become broader. Researchers then began looking at integrated management strategies for bark beetle-host systems (McGregor and Cole 1985, Waters et al. 1985). Syntheses of the state of knowledge of the cause and effect role of bark beetles and bark beetle management in the interior northwest have been published (Gast et al. 1991, Filip et al.1996, Hayes and Daterman 2001). By the 1990s, ecosystem and landscape management demands called for different analytical systems that examined multiple resources and could handle greater complexity and scale. Landscape simulation models provide a means of projecting long-term and large scale changes from succession, management, and disturbance. Using many of the same attributes of the early classification systems, models such as the Douglas-fir beetle impact model (Marsden et al. 1997) and the Western Pine Beetle Model (Beukema et al. 1997) were developed as extensions to the Forest Vegetation Simulator, which when integrated with other submodels allows simulations of multiple processes (e.g., Ager et al. 2007, McMahan et al. 2008). At larger scales, coarser grain models, such as state and transition models, have been used to examine multiple resource variables along with bark beetles and other insects (Hessburg et al. 1999, Barbour et al. 2007, Hemstrom et al. 2007). Limited research has directly addressed societal reactions to bark beetle outbreaks (e.g., Flint 2006). Additional research and improvements in landscape simulation models that include socioeconomic components and permit robust analysis of tradeoffs for management options including no treatment alternatives are needed. This is an area of high priority research for the WBBRG. For a more detailed description of this topic, see Progar and others in this Proceedings.

Bark beetle and fire interactions—It is generally acknowledged that historically across western landscapes, particularly in dry interior forests, disturbance agents including wildland fire and insects influenced successional processes (Agee 2003). Fire suppression over the past 100 years has changed both the frequency and severity of wildfire and insect outbreaks (Hessburg et al. 1994). Stimulated in part by large fires at the beginning of this century, researchers have increasingly placed more emphasis on the apparent reciprocal and sometimes synergistic association between fire and bark beetles; previous research efforts are reviewed by McCullough et al. (1998) and more recently by Parker et al. (2006). Recent studies have begun to examine functional and numerical interactions between bark beetles and fire at the tree and stand level (e.g., Hood and Bentz 2007), and the relationship between beetle outbreak and fuel dynamics (e.g., Jenkins et al. 2008). Few quantitative studies have been carried out that consider the spatiotemporal dynamics of wildfire and bark beetle outbreaks at the landscape scale. Somewhat conflicting results to date are indications that the mechanisms are complex, particularly over time and at large spatial scales. Given the number of forested acres affected by insects and wildland fire each year, it's clear why the interactions between fire and bark beetles continues to be an area of high priority research for the WBBRG. For a more detailed description of this topic, see Gibson and Negrón in this Proceedings.

Responses of bark beetle systems to climate change—Investigations of the role of historical or natural range of variation of bark beetles are limited, but are important for understanding when and how they function as natural disturbance agents in forest ecosystems. Understanding how landscapes respond over time to perturbations including climate change is key to the development of effective forest management strategies for the future. The response of beetles, their common associates and hosts is an area of active investigation by the FS R&D researchers and one of WBBRG's priority areas. Research is ongoing at the individual and mechanistic level (e.g., Bentz and Mullen 1999, Hansen et al. 2002, Six and Bentz 2007), as well as at the population and landscape level (e.g., Logan and Powell 2001, Logan et al. 2003, Regniere and Bentz 2006). For a more detailed discussion of this topic, see Lundquist and Bentz in this Proceedings.

Chemical and semiochemical-based management tactics—Research in the area of direct control of bark beetles and use of pesticides began in the mid-1900s and continues today (e.g., Negrón et al. 2001, Fettig et al. 2006). Direct control measures often have limited but important applications, particularly in high value areas. For example, research on viable replacements for carbaryl (Hastings et al. 2001), one of the most effective treatments for individual trees against attack by many bark beetles (Fettig et al. 2006), is likely to continue. Others have determined the amount of drift that occurs during these treatments and used this information to determine the potential risk that drift poses to fish and other taxa in nearby aquatic systems, a primary concern when treating trees in campgrounds in the West (Fettig et al. 2008).

Behavior- or semiochemical research has been a strong component and focus of research in FS R&D in the West since the early studies in the late 60s and 70s. New technologies in both experimental exploration and application continue to make this a productive area of research and development for detection and suppression tactics. Using techniques that are crude by today's standards, early researchers succeeded in identifying the attractant or aggregant, and anti-aggregant pheromones, along with synergistic compounds, produced by many major bark beetles (e.g., Furniss et al. 1972). Attractants have long been used in trapping technology for detection and monitoring. The relationship between trap captures and population dynamics and more specifically, levels of bark beetle-caused tree mortality in a given area remains an area of active research (e.g., Bentz 2006, Hansen et al. 2006). The role of host tree physiology and host-produced volatiles is also an area of ongoing research (e.g., Joseph et al. 2001, Kelsey and Joseph 2001, 2003, Kelsey and Manter 2004, Manter and Kesley 2008).

Similarly, development of semiochemical-based suppression tactics has been an active and effective area of research for FS R&D in the West. Development of individual tree to area-wide protection from infestation for some bark beetles such as Douglas-fir beetle (e.g., Ross and Daterman 1994, 1995, 1997) and mountain pine beetle (e.g., Progar 2005, Gillette et al. 2006) represent important tools for managers. Further improvements and development of similar tools for protecting single-tree to large-scale areas from other bark beetle species are needed, particularly for high value resources. This area continues to be an area of high priority research for the WBBRG. For a more detailed discussion of this research topic see Gillette and Munson in this Proceedings.

Detection, monitoring, and management of bark and woodboring invasives—Many of the same technologies used for native species are being applied to the research and development of detection and mitigation tools for non-native invasive bark and woodboring beetles. Non-native insects are not new to the conifer forests of the western U.S., but represent an increasing threat as global trade and human traffic brings increased opportunities for importation and exchange. Between 1985 and 1998, approximately 90% of the non-native insects intercepted on wood materials were Coleoptera, and of those introduced beetles, over 50% belong to the bark and woodboring Scolytinae (Haack and Cavey 2000, Haack 2006). Many of the most noteworthy introductions have been in the eastern U.S. (e.g., pine shoot borer, Asian longhorn beetle, emerald ash borer); however, by direct importation or spread from elsewhere within North America, the number of invaders continues to grow in the West (e.g., Lee et al. 2007). Surveys conducted only a few years apart reveal new non-native woodborer records for the Pacific Northwest and North America (Mudge et al. 2001, LaBonte et al. 2005). In the past, western forest entomologists have studied a number of invaders, particularly defoliating Lepidoptera (Hayes and Ragenovich 2001). One with potential for changing forest composition was the larch casebearer which spread from the East. A classic biological control treatment was developed by an FS R&D research entomologist for this defoliator (Ryan 1997). It is a textbook example; there have been no documented non-target effects of the non-native parasitoids released to control this invader, the control has been maintained for over a decade, and it appears to be self-sustaining. Detection, monitoring, and management for invasive bark and woodboring beetles is an area of current research (e.g., Negrón et al. 2005, Johnson et al. 2008, Lee et al. 2008, Liu et al. 2008) and a high priority area for the WBBRG. For a more detailed discussion of this research topic see Seybold and Downing in this Proceedings.

Working with our FHP partners and others, the WBBRG seeks to continue this legacy of relevant research, delivery and partnership. Exemplying this spirit and representative of our mutual goals to work cooperatively and communicate with stakeholders, each of the informative papers in this Proceedings is a WBBRG and FHP collaboration.

Acknowledgments

We thank John Lundquist (Forest Health Protection and Pacific Northwest Research Station, USDA Forest Service, Anchorage, AK) for organizing and inviting us to participate in this symposium. We also thank our colleagues who provided photographs used in our presentation and this paper; photographs obtained from the PNW historic photo collection housed at the La Grande Forestry and Range Sciences Laboratory were also used in the presentation. Christopher J. Fettig (Pacific Southwest Research Station, USDA Forest Service, Davis, CA), Rick Kelsey (Pacific Northwest Research Station, Corvallis, OR), and Lia Spiegel (Blue Mountains Pest Management Service Center, Wallow-Whitman NF, USDA Forest Service, La Grande, OR) for reviewing earlier versions of this manuscript.

Literature Cited

Agee, J.K. 2003. Historical range of variability in eastern Cascades forests, Washington USA. Landscape Ecology. 18: 725–740.

Ager, A.A.; MacMahan, A.; Hayes, J.L.; Smith, E.L. 2007. Long term simulation of potential bark beetle impacts and stand-level fire behavior on thinned versus unmanaged forest landscapes. Landscape and Urban Planning. 80(3): 301–311.

Barbour, R.J.; Hemstrom, M.A.; Hayes, J.L. 2007. The Interior Northwest Landscape Analysis System: a step toward understanding integrated landscape management planning. Landscape and Urban Planning. 80(3): 333–344.

Bentz, B.J. 2006. Mountain pine beetle population sampling: inferences from Lindgren pheromone traps and tree emergence cages. Canadian Journal of Forest Research. 36(2): 351–360.

Bentz, B.J.; Mullens, D.E. 1999. Ecology of mountain pine beetle cold hardening in the Intermountain West. Environmental Entomology. 28: 577–587.

Berg, E.E.; Henry, J.D.; Fastiec, C.L.; De Volder, A.D.; Matsuokae, S.M. 2006. Spruce beetle outbreaks on the Kenai Peninsula, Alaska, and Kluane National Park and Reserve, Yukon Territory: Relationship to summer temperatures and regional differences in disturbance regimes. Forest Ecology and Management. 227(3): 219–232.

Beukema, S.J.; Greenough, J.A.; Robinson, D.C.E.; Kurz, W.A.; Smith, E.L.; Eav, B.B. 1997. The Westwide Pine Beetle Model: a spatially-explicit contagion model. In: Teck, R.; Moeur, M.; Adams, J., comps. Proceedings of the Forest Vegetation Simulator Conference. Gen. Tech. Rep. INT-GTR-373. Ogden, UT: U.S. Department of Agriculture, Forest Service, Intermountain Research Station: 126–130.

Breshears, D.D.; Cobb, N.S.; Rich, P.M.; Price, K.P.; Allen, C.D.; Balice, R.G.; Romme, W.H.; Kastens, J.H.; Floyd, M.L.; Belnap, J.; Anderson, J.J.; Myers, O.B.; Meyer, C.W. 2005. Regional vegetation die-off in response to a global-change-type drought. Proceedings of the National Academy of Sciences. 102(42): 15144–15148.

Cole, W. E.; Amman, G. D. 1980. Mountain pine beetle dynamics in lodgepole pine forests. Part 1: Course of an infestation. Gen. Tech. Rep. INT-GTR-89. Ogden, UT: U.S. Department of Agriculture, Forest Service, Intermountain Forest and Range Experiment Station. 56 p.

Fettig C.J.; Allen K.K.; Borys, R.R.; Christopherson, J.; Dabney, C.P.; Eager, T.J.; Gibson, K.E.; Hebertson, E.G.; Long, D.F.; Munson, A.S.; Shea, P.J.; Smith, S.L.; Haverty, M.I. 2006. Effectiveness of bifenthrin (Onyx) and carbaryl (Sevin SL) for protecting individual tree, high-value conifers from bark beetle attack (Coleoptera: Curculionidae: Scolytinae) in the Western United States. Journal of Economic Entomology. 99: 1691–1698.

Fettig, C.J.; Klepzig, K.D.; Billings, R.F.; Munson, A.S.; Nebeker, T.E.; Negrón, J.F.; Nowak, J.T. 2007. The effectiveness of vegetation management practices for prevention and control of bark beetle outbreaks in coniferous forests of the western and southern United States. Forest Ecology and Management. 238: 34–53.

Fettig, C.J.; Munson, A.S.; McKelvey, S.R.; Bush, P.B.; Borys, R.R. 2008. Spray deposition from ground-based applications of carbaryl to protect individual trees from bark beetle attack. Journal of Environmental Quality. 37: 1170-1179.

Filip, G.M.; Torgersen, T.R.; Parks, C.A.; Mason, R.R.; Wickman, B. 1996. Insects and disease factors in the Blue Mountains. In: Jaindl, R.G.; Quigley, T.M., eds., Search for a Solution: Sustaining the Land, People, and Economy of the Blue Mountains. American Forests, Washington, DC: 169–202.

Flint, C.G. 2006. Community perspectives on spruce beetle impacts on the Kenai Peninsula, Alaska. Forest Ecology and Management. 227: 207–218.

Furniss, M.M. 1979. An annotated bibliography of the Douglas-fir beetle (*Dendroctonus pseudotsugae* Hopkins). Gen. Tech. Rep. INT-GTR-48. Ogden, UT: U.S. Department of Agriculture, Forest Service Intermountain Forest and Range Experiment Station. 40 p.

Furniss, M.M.; Kline, L.N.; Schmitz, R.F.; Rudinsky, J.A. 1972. Tests of three pheromones to induce or disrupt aggregation of Douglas-fir beetles (Coleoptera: Scolytidae) on live trees. Annals of the Entomological Society of America. 65: 1227–1232.

Gast, W.R. Jr.; Scott, D.W.; Schmitt, C.; Clemens, D.; Howes, S.; Johnson, C.G. Jr.; Mason, R.; Mohr, F.; Clapp, R.A. Jr. 1991. Blue Mountains forest health report: new perspectives in forest health. U.S. Department of Agriculture, Forest Service, Malheur, Umatilla, and Wallowa-Whitman National Forests. [Unconventional pagination].

Gibson, K.; Negrón J.F. 2009. Fire and bark beetle interactions. In: Hayes, J.L.; Lundquist, J.E., comps. Western Bark Beetle Research Group—a unique collaboration with Forest Health Protection symposium, Society of American Foresters Conference, 23-28 October 2007, Portland, OR. Gen. Tech. Rep. PNW-GTR-784, Portland, OR: U.S. Department of Agriculture, Forest Service, Pacific Northwest Research Station: 51–69.

Gillette, N.E.; Munson A.S. 2009. Semiochemical sabotage: behavioral chemicals for protection of western conifers from bark beetle. In: Hayes, J.L.; Lundquist, J.E., comps. Western Bark Beetle Research Group—a unique collaboration with Forest Health Protection symposium, Society of American Foresters Conference, 23-28 October 2007, Portland, OR. Gen. Tech. Rep. PNW-GTR-784, Portland, OR: U.S. Department of Agriculture, Forest Service, Pacific Northwest Research Station: 85–109.

Gillette, N.E.; Stein, J.D.; Owen, D.R.; Webster, J.N.; Fiddler, G.O. 2006. Verbenone-releasing flakes protect individual *Pinus contorta* trees from attack by *Dendroctonus ponderosae* and *Dendroctonus valens* (Coleoptera: Curculionidae: Scolytinae). Agricultural and Forest Entomology. 8: 243–251.

Haack R.A.; Cavey, J.F. 2000. Insects intercepted on solid wood packing materials at United States ports-of-entry: 1985–1998. In: Quarantine pest, risk for the forestry sector and their effects on foreign trade. Proceedings on CD-ROM of Silvotecna 14, 27–28 June 2000, Concepcion, Chile. CORMA, Concepcion, Chile: 1–16.

Haack, R.A. 2006. Exotic bark- and wood-boring Coleoptera in the United States: recent establishment and interceptions. Canadian Journal of Forest Research. 36: 269–288.

Hansen. E.M.; Bentz, B.J.; Turner, D.L. 2001. Temperature-based model for predicting univoltine brood proportions in spruce beetle (Coleoptera: Scolytidae). Canadian Entomologist. 133(6): 827–841.

Hansen, E.M.; Bentz, B.J.; Munson, A.S. 2006. Evaluation of funnel traps for estimating tree mortality and associated population phase of spruce beetle in Utah. Canadian Journal of Forest Research. 36(10): 2574–2584.

Hastings, F.L.; Holsten, E.H.; Shea, P.J.; Werner, R.A. 2001. Carbaryl: a review of its use against bark beetles in coniferous forests of North America. Environmental Entomology. 30: 803–810.

Hayes, J.L.; Daterman, G.E. 2001. Bark beetles (Scolytidae) in Eastern Oregon and Washington. Northwest Science. 75(Special issue): 21–30.

Hessburg, P.F.; Mitchell, G.R.; Filip, G.M. 1994. Historical and current roles of insects and pathogens in eastern Oregon and Washington forested landscapes. Gen. Tech. Rep. PNW-GTR-327. Portland, OR: U.S. Department of Agriculture, Forest Service, Pacific Northwest Research Station. 72 p.

Hessburg, P.F.;Smith, B.G.; Miller, C.A.; Kreiter, S.D.; Slater, R.B. 1999. Modeling change in potential landscape vulnerability to forest insect and pathogen disturbances: methods for forested subwatersheds sampled in the midscale interior Columbia River basin assessment. Gen. Tech. Rep. PNW-GTR-454. Portland, OR: U.S. Department of Agriculture, Forest Service. Pacific Northwest Research Station, 56 p.

Hood, S.; Bentz, B.J. 2007. Predicting postfire Douglas-fir beetle attacks and tree mortality in the Northern Rocky Mountains. Canadian Journal of Forest Research. 37: 1058–1069.

Jenkins, M.J.; Hebertson, E.; Page, W.; Jorgersen, C.A. 2008. Bark beetles, fuels, fire and implications for forest management in the Intermountain West. Forest Ecology and Management. 254: 16–34.

Joseph, G.; Kelsey, R.G.; Peck, R.W.; Niwa, C.G. 2001. Response of some scolytids and their predators to ethanol and 4-allylanisole in pine forests of central Oregon. Journal of Chemical Ecology. 27(4): 697–715.

Juday, G.P. 1998. Spruce beetles, budworms, and climate warming. Global Glimpses. 6(1) http://www.cgc.uaf.edu/Newsletter/gg6_1/beetles.html. (3 December 2008).

Keen, F.P. 1943. Ponderosa pine tree classes redefined. Journal of Forestry. 41: 249–253.

Kelsey, R.G.; Joseph, G. 2001. Attraction of *Scolytus unispinosus* bark beetles to ethanol in water-stressed Douglas-fir branches. 2001. Forest Ecology and Management. 144(1-3): 229–238.

Kelsey, R.G.; Joseph, G. 2003. Ethanol in ponderosa pine as an indicator of physiological injury from fire and its relationship to secondary beetles. Canadian Journal of Forest Research. 33: 870–884.

Kelsey, R.G.; Manter, D.K. 2004. Effect of Swiss needle cast on Douglas-fir stem ethanol and monoterpene concentrations, oleoresin flow, and host selection by the Douglas-fir beetle Forest Ecology and Management. 190(2–3): 241–253.

LaBonte, J.R.; Mudge, A.D.; Johnson, K.L. 2005. Nonindigenous woodboring Coleoptera (Cerambycidae, Curculionidae: Scolytinae) new to Oregon and Washington, 1999–2002: Consequences of the intracontinental movement of raw wood products and solid wood packing materials. Proceedings of the Entomological Society of Washington. 107(3): 554–564.

Lee, J.C.: Haack, R.A.; Negrón, J.F.; Witcosky, J.J.; Seybold, S.J. 2007. Invasive bark beetles, Forest Insect & Disease Leaflet 176. Washington DC: U.S. Department of Agriculture, Forest Service.

Lee, J.C.; Flint, M.L.; Seybold, S.J. 2008. Suitability of pines and other conifers as hosts for the invasive Mediterranean pine engraver (Coleoptera: Scolytidae) in North America. Journal of Economic Entomology. 101: 829–837.

Lessard, G.; Hansen, M.; Eilers, M.L. 1986. Annotated bibliography: mountain pine beetle, *Dendroctonus ponderosae*, Hopkins. Forest Service Technical Report R2-35. Lakewood, CO: U.S. Department of Agriculture, Rocky Mountain Region, Timber, Forest Pest, and Cooperative Forestry Management. 157 p.

Linton, D.A.; Safranyik, L. 1987. The spruce beetle *Dendroctonus rufipennis* (Kirby): an annotated bibliography 1885–1987. Canadian Forestry Service Information Report BC-X-298. Pacific Forestry Center, Hull, Quebec. 38 p.

Liu, D.-G.; Flint, M.L.; Seybold, S.J. 2008. A secondary sexual character in the redhaired pine bark beetle, *Hylurgus ligniperda* Fabricius (Coleoptera: Scolytidae). The Pan-Pacific Entomologist. 84: 26–28.

Logan, J.A.; Powell, J.A. 2001. Ghost forests, global warming, and the mountain pine beetle. American Entomologist. 47: 160–73.

Logan, J.A.; Regniere, J.; Powell, J.A. 2003. Assessing the impacts of global warming on forest pest dynamics. Frontiers in Ecology and Environment. 1: 130–137.

Lundquist, J.E.; Bentz B.J. 2009. Bark beetles in a changing climate. In: Hayes, J.L.; Lundquist, J.E., comps. Western Bark Beetle Research Group—a unique collaboration with Forest Health Protection symposium, Society of American Foresters Conference, 23–28 October 2007, Portland, OR. Gen. Tech. Rep. PNW-GTR-784, Portland, OR: U.S. Department of Agriculture, Forest Service, Pacific Northwest Research Station: 39–49.

Manter, D.K.; Kelsey, R.G. 2008. Ethanol accumulation in drought-stressed conifer seedlings. International Journal of Plant Science. 169: 361–369.

Marsden, M.A.; Eav, B.B.; Thompson, M.K. 1994. User's guide to the Douglas-fir beetle impact model. Gen. Tech. Rep. RMRS-GTR-250. Fort Collins, CO: U.S. Department of Agriculture, Forest Service, Rocky Mountain Research Station. 9 p.

McMahan, A.J.; Ager, A.A.; Maffei, H.; Hayes, J.L.; Smith, E.L. 2008. Modeling bark beetles and fuels on landscapes: a demonstration of ArcFuels, and a discussion of modeling pursuits. In: Havis, R.N.; Crookston, N.L., comps. Third Forest Vegetation Simulator Conference. RMRS-P-54. Fort Collins, CO: U.S. Department of Agriculture, Forest Service, Rocky Mountain Research Station: 40–52.

McCullough, D.G.; Werner, R.A.; Newmann, D. 1998. Fire and insects in northern and boreal forest ecosystems of North America. Annual Review of Entomology. 43: 107–127.

McGregor, M.D.; Cole, D.M. 1985. Integrating management strategies for the mountain pine beetle with multiple-resource management of lodgepole pine forests. Gen. Tech. Rep. INT-GTR-174. Ogden, UT: U.S. Department of Agriculture, Forest Service, Intermountain Forest and Range Experiment Station. 69 p.

McMillin, J.D.; Fettig, C.J. 2009. Bark beetle responses to vegetation management treatments. In: Hayes, J.L.; Lundquist, J.E., comps. Western Bark Beetle Research Group—a unique collaboration with Forest Health Protection symposium, Society of American Foresters Conference, 23–28 October 2007, Portland, OR. Gen. Tech. Rep. PNW-GTR-784, Portland, OR: U.S. Department of Agriculture, Forest Service, Pacific Northwest Research Station: 25–38.

Miller, J.M.; Keen, F.P. 1960. Biology and control of the western pine beetle. U.S. Department of Agriculture, Forest Service Miscellaneous Publication 800. Washington, DC. 381 p.

Mudge, A.D.; LaBonte, J.R; Johnson, K.L.; LaGasa, E.H. 2001. Exotic woodboring Coleoptera (Micromalthidae, Scolytidae) and Hymenoptera (Xiphydriidae) new to Oregon and Washington. Proceedings of the Entomological Society of Washington. 103(4): 1011–1019.

Negrón, J.F. 1997. Estimating probabilities of infestation and extent of damage by roundheaded pine beetles in ponderosa pine in the Sacramento Mountains, New Mexico. Canadian Journal of Forest Research. 27: 1936–1945.

Negrón, J.F. 1998. Probability of infestation and extent of mortality associated with the Douglas-fir beetle in the Colorado Front Range. Forest Ecology and Management. 107: 71–85.

Negrón, J.F.; Popp, J.B. 2004. Probability of ponderosa pine infestation by mountain pine beetle in the Colorado Front Range. Forest Ecology and Management. 191: 17–27.

Negrón, J.F.; Sheppard, W.D.; Mata, S.A.; Popp, J.B.; Asherin, L.A.; Schoettle, A.W.; Schmid, J.M.; Leatherman, D.A. 2001. Solar treatments for reducing survival of mountain pine beetle in infested ponderosa and lodgepole pine logs. RMRS-RP-30. Fort Collins, CO: U.S. Department of Agriculture, Forest Service, Rocky Mountain Research Station. 11 p.

Negrón, J.F.; Witcosky, J.J.; Cain, R.J; LaBonte, J.R.; Duerr, D.A., II; McElwey, S.J.; Lee, J.C; Seybold, S.J. 2005. The banded elm bark beetle: a new threat to elms in North America. American Entomologist. 51(2): 84–94.

Negrón, J.F.; Allen K.; Cook, B.; Withrow, J.R., Jr. 2008a. Susceptibility of ponderosa pine, *Pinus ponderosa* (Dougl. Ex. Laws.), to mountain pine beetle, *Dendroctonus ponderosae* Hopkins, attack in uneven-age stands in the Black Hills of South Dakota and Wyoming, USA. Forest Ecology and Management. 254: 327–334.

Negrón, J.F.; Bentz, B.J.; Fettig, C.J.; Gillette, N.E.; Hansen, E.M.; Hayes, J.L.; Kelsey, R.G.; Lundquist, J.E.; Lynch, A.M.; Progar, R.A.; Seybold, S.J. 2008b. U.S. Forest Service bark beetle research in the western United States: Looking toward the future. Journal of Forestry. 106: 325–331.

Parker, T.J.; Clancey, K.M.; Mathiasen, R.L. 2006. Interactions among fire, insects and pathogens in coniferous forests of the interior western United States and Canada. Agricultural and Forest Entomology. 8: 167–189.

Progar, R.A. 2005. Five-year operational trial of verbenone to deter mountain pine beetle (*Dendroctonus ponderosae*; Coleoptera: Scolytidae) attack on lodgepole pine (*Pinus contorta*). Environment Entomology. 34: 1402–1407.

Progar, R.A.; Eglitis, A.; Lundquist, J.E. 2009. Some ecological, economic, and social consequences of bark beetle infestations. In: Hayes, J.L.; Lundquist, J.E., comps. Western Bark Beetle Research Group—a unique collaboration with Forest Health Protection symposium, Society of American Foresters Conference, 23–28 October 2007, Portland, OR. Gen. Tech. Rep. PNW-GTR-784. Portland, OR: U.S. Department of Agriculture, Forest Service, Pacific Northwest Research Station: 71–83.

Ryan, R.B. 1997. Before and after evaluation of biological control of the larch casebearer (Lepidoptera: Coleophoridae) in the Blue Mountains of Oregon and Washington, 1972-1995. Environmental Entomology. 26(3): 703–715.

Regniere J.; Bentz B.J. 2007. Modeling cold tolerance in the mountain pine beetle, *Dendroctonus ponderosae*. Journal of Insect Physiology. 53(6): 559–572.

Ross, D.W.; Daterman, G.E. 1994. Reduction of Douglas-fir beetle infestation in high risk stands by antiaggregation and aggregation pheromones. Canadian Journal of Forest Research. 24: 2184–2190.

Ross, D.W.; Daterman, G.E. 1995. Efficacy of an antiaggregation pheromone for reducing Douglas-fir beetle, *Dendroctonus pseudotsugae* Hopkins (Coleoptera: Scolytidae), infestation in high risk stands. Canadian Entomologist. 127: 805–811.

Ross, D.W.; Daterman, G.E. 1997. Integrating pheromone and silvicultural methods for managing the Douglas-fir beetle. In: Gregoire, J.C.; Liebhold, A.M.; Stephen, F.M.; Day, K.R.; Salom, S.M., eds. Proceedings—Integrating cultural tactics into the management of bark beetles and reforestation pests. Gen. Tech. Rep. NE-GTR-236. Radnor, PA: U.S. Department of Agriculture, Forest Service, Northeastern Research Station: 135–145.

Schmid, J.M.; Frye, R.H. 1976. Stand ratings for spruce beetles. Res. Note RM-309. Fort Collins, CO: U.S. Department of Agriculture, Forest Service, Rocky Mountain Forest and Range Experiment Station, 4 p.

Seybold, S.J.; Downing, M. 2009. What risk do invasive bark beetles and woodborers pose to forests of the western U.S?: a case study of the Mediterranean pine engraver, *Orthotomicus erosus*. In: Hayes, J.L.; Lundquist, J.E., comps. Western Bark Beetle Research Group— a unique collaboration with Forest Health Protection symposium, Society of American Foresters Conference, 23–28 October 2007, Portland, OR. Gen. Tech. Rep. PNW-GTR-784. Portland, OR: U.S. Department of Agriculture, Forest Service, Pacific Northwest Research Station: 111–134.

Six, D.L.; Bentz, B.J. 2007. Temperature determines symbiont abundance in a multipartite bark beetle-fungus ectosymbiosis. Microbial Ecology. 54(1): 112–118.

Stevens, R.T.; McCambridge, W.F.; Edminster, C.B. 1980. Risk rating guide for mountain pine beetle in Black Hills ponderosa pine. Res. Note RM-385. Fort Collins, CO: U.S. Department of Agriculture, Forest Service, Rocky Mountain Forest and Range Experiment Station. 2 p.

U.S. Department of Agriculture, Forest Service. 2005. Forest insect and disease conditions in the United States 2004. Forest Health Protection. Washington DC, 142 p.

Waters, W.W.; Stark, R.W.; Wood, D.L. 1985. Integrated Pest Management for Pine-Bark Beetle Ecosystems. New York: John Wiley & Sons.

Western Forestry Leadership Coalition 2007. Western bark beetle assessment: a framework for cooperative forest stewardship. http://www.wflccenter.org/infomaterials/reports.php. (3 December 2008).

Wickman, B.E. 2005. Harry E. Burke and John M. Miller, pioneers in western forest entomology. Gen. Tech. Rep. PNW-GTR-638. Portland, OR: U.S. Department of Agriculture, Forest Service, Pacific Northwest Research Station. 163 p.

Wickman, B.E.; Torgersen, T.R.; Furniss, M.M. 2002. Photographic images and history of forest insect investigations on the Pacific slope, 1903–1953. Part 2 Oregon and Washington. American Entomologist. 178–185.

Bark Beetle Responses to Vegetation Management Treatments[1]

Joel D. McMillin and Christopher J. Fettig[2]

Abstract

Native tree-killing bark beetles (Coleoptera: Curculionidae, Scolytinae) are a natural component of forest ecosystems. Eradication is neither possible nor desirable and periodic outbreaks will occur as long as susceptible forests and favorable climatic conditions co-exist. Recent changes in forest structure and tree composition by natural processes and management practices have led to increased competition among trees for water, nutrients and growing space thereby increasing tree stress. As trees become stressed, their insect resistance mechanisms are compromised and thus they become more susceptible to bark beetle attack. In this presentation, we reviewed tree and stand factors associated with bark beetle infestations and analyzed the effectiveness of vegetation management practices for mitigating the negative impacts of bark beetles on forest ecosystems. We described the current state of our knowledge and practical application of this knowledge; identified future research needs required to make informed decisions on proposed silvicultural treatments; and discussed ongoing research efforts led by the Western Bark Beetle Research Group. Our discussion concentrated on pine-dominated systems in the western US.

Keywords: Silviculture, thinning, prescribed fire, bark beetles, ponderosa pine.

---

[1] The genesis of this manuscript was a presentation by the authors at the Western Bark Beetle Research Group—A Unique Collaboration with Forest Health Protection Symposium, Society of American Foresters Conference, 23–28 October 2007, Portland, OR.

[2] **Joel D. McMillin** is an Entomologist, Forest Health Protection, USDA Forest Service, 2500 South Pine Knoll Drive, Flagstaff, AZ 86001; email: jmcmillin@fs.fed.us. **Christopher J. Fettig** is a Research Entomologist, Pacific Southwest Research Station, USDA Forest Service, 1731 Research Park Drive, Davis, CA 95616; email: cfettig@fs.fed.us.

Introduction

Bark beetles (Coleoptera: Curculionidae, Scolytinae), a large and diverse group of insects consisting of approximately 550 species in North America (Wood 1982), are commonly recognized as the most important mortality agent in coniferous forests (Furniss and Carolin 1977). Most bark beetles feed on the phloem tissue of woody plants and often directly kill the host influencing forest ecosystem structure and function by regulating certain aspects of primary production, nutrient cycling, ecological succession and the size, distribution and abundance of forest trees (Mattson 1977, Mattson and Addy 1975, Mattson et al. 1996). Attacks reduce tree growth and hasten decline, mortality and subsequent replacement by other tree species. Severe infestations may impact timber and fiber production, water quality and quantity, fish and wildlife populations, recreation, grazing capacity, biodiversity, endangered species, real estate values and cultural resources in a variety of ways.

Individual trees utilize growth factors until one or more factors become limiting (Oliver and Larson 1996). Therefore, a forest contains a certain amount of intangible growing space, which varies spatially and temporally. Disturbances can make growing space available to some tree species at the expense of others (e.g., selective herbivory), or alter the amount of growing space available to all trees (e.g., prolonged drought) (Fettig et al. 2007). As growing space diminishes, a tree's photosynthates are allocated to different uses in an order of priorities (Oliver and Larson 1996): (1) maintenance respiration (Kramer and Kozlowski 1979), (2) production of fine roots (Fogel and Hunt 1979), (3) reproduction (Eis et al. 1965), (4) primary (height) growth (Oliver and Larson 1996), (5) xylem (diameter) growth (Waring and Schlesinger 1985), and (6) insect and disease resistance mechanisms (Mitchell et al. 1983). This hierarchy is not absolute, but is often used to illustrate how production of insect resistance mechanisms may be compromised when growing space becomes limited by one or more factors (Fettig et al. 2007).

In order to reproduce, bark beetles must successfully locate and colonize suitable hosts. Once identified, using a variety of behavioral modalities, host colonization begins with the biting process. Given the cues received during this process and other factors, such as the beetle's internal physiology (Wallin and Raffa 2000), the host is either rejected or accepted. If the host is rejected, the beetle takes flight presumably in search of another host. If the host is accepted, colonization in the case of living hosts requires overcoming tree defenses that consist of anatomical and chemical components that are both constitutive and inducible (Franceschi et al. 2005). This can only be accomplished by recruitment of a critical minimum number of beetles, which varies with changes in host vigor (Berryman 1982). Most coniferous species, particularly pines, have a well-defined resin duct system, which is capable of mobilizing large amounts of oleoresin upon wounding and often drowns or encapsulates attacking beetles.

Factors such as stand density, basal area or stand density index, tree diameter and host density are consistently identified as primary attributes associated with bark beetle infestations. Therefore, efforts to prevent undesirable levels of bark beetle-caused tree mortality must change stand susceptibility through reductions in tree competition and/or changes in tree species composition.

Bark Beetle Responses to Vegetation Management Treatments
Based on a comprehensive review of empirical and anecdotal evidence concerning the effects of thinning and other vegetation management practices on host susceptibility and subsequent bark beetle infestation, Fettig et al. (2007) developed seven primary conclusions. These are paraphrased below and supplemented with additional supporting information.

1. Bark beetles causing the majority of conifer mortality in the US are native insects and an integral component of forest ecosystems. As such, eradication is neither possible nor desirable. Although bark beetles are native to conifer forests of the western US, conditions of many forest types have changed substantially over the past century (Cocke et al. 2005), resulting in increased inter-tree competition and subsequent landscape level outbreaks (USDA Forest Service 2005). Changing forest stand and tree conditions through vegetation management would sensibly decrease susceptibility to bark beetle-caused impacts.

2. Forested landscapes that contain little heterogeneity promote the creation of large contiguous areas susceptible to insect outbreaks. For example, the extensive mountain pine beetle, *Dendroctonus ponderosae* Hopkins, outbreak in British Columbia, Canada may be due in part to homogenization of forest stands over large geographic areas. In the early 1900s, ~17 percent of lodgepole pine, *Pinus contorta* Dougl. ex Loud., forests were in age classes susceptible to mountain pine beetle infestation, while today >50 percent of forests meet this classification (Taylor and Carroll 2004). When developing vegetation management strategies for bark beetles, susceptibility needs to be considered at both stand and landscape levels. Typically, the later is often not adequately addressed.

3. Although an extensive body of research exists describing relationships among stand conditions, vegetation management practices, and host susceptibility for several bark beetle species (e.g., mountain pine beetle), we still have research gaps for some cover types and common bark beetle species (e.g., bark beetles attacking true fir species). McMillin et al. (2003) related the extent of subalpine fir, *Abies lasiocarpa* (Hook.) Nutt., mortality caused by western balsam bark beetle, *Dryocoetes confusus* Swaine, to forest conditions in north-central Wyoming. Significant positive linear relationships were found between amount of fir mortality and percentage of subalpine fir trees, subalpine fir basal area, and subalpine fir stand density index. However, additional studies are required to more fully understand factors associated with bark beetle infestations in true fir forests, and to develop silvicultural prescriptions to minimize undesirable levels of western balsam bark beetle-caused tree mortality.

4. Bark beetle infestations are consistently associated with certain forest stand and site conditions, such as tree density, basal area, stand density index, and site quality index. These findings have implications for developing vegetation management strategies.

Although not all studies examining the effects of thinning have demonstrated significant treatment effects, no studies have shown that thinning resulted in significant increases in the amount of *Dendroctonus*-caused tree mortality. Furthermore, vegetation management treatments can have direct and indirect societal benefits in addition to reducing tree losses associated with bark beetle infestations. For example, thinning can redistribute growing space to desirable trees, utilize anticipated mortality resulting from stem exclusion, encourage regeneration, create early cash flows, and reduce risks associated with fire and diseases.

5. Several bark beetles are attracted to thinning residues (slash), most notably several species in the genus Ips *(Livingston 1979, Parker 1991).* The most damaging effects occur when fresh slash and weakened trees are present in an area for two or more years (Parker 1991). However, impacts caused by bark beetles infesting thinning residues can be minimized through the use of published guidelines (DeGomez et al. 2008, Kegley et al. 1997, Parker 1991), which include information regarding the timing of thinning, slash size, removal of thinning residues, and appropriate treatment of slash by burning, chipping, or burying (see "*Ips*-n-chips" section below for more on slash management and bark beetles).

6. Sublethal heating of critical plant tissue can stress trees and increase their susceptibility to bark beetle attack. Prescribed fires are increasingly being implemented to reduce the risk of catastrophic wildland fires (Agee and Skinner 2005); however, there is the potential for unintended increases in bark beetle activity to occur following relatively low-intensity prescribed fires (Parker et al. 2006). For example, Breece et al. (2008) found a significantly greater proportion of ponderosa pine, *P. ponderosa* Dougl. ex Laws., trees attacked by bark beetles in stands that were prescribed burned (13%) than in paired unburned stands (1.5%) at sites in Arizona and New Mexico. However, the authors stated that relatively small increases in tree mortality should be acceptable to many forest managers given the effects of such fuels management treatments on reducing surface fuel loads and the risk of severe wildfire.

7. The effectiveness of direct control techniques varies among bark beetle species. For example, direct control treatments (i.e., cut-and-remove, cut-and-leave) can be effective for managing southern pine beetle, *D. frontalis* Zimmermann, infestations because of its unique life cycle and attack behavior (Billings 1995). In general, these treatments are not as effective for management of bark beetle species in the western US, especially once an epidemic population phase has been reached. Most effective direct control treatments in the West are those that target increasing, but localized populations and those that are in response to discrete disturbance events (e.g., windthrow, mixed-severity fire).

Vegetation treatments currently implemented in southwestern ponderosa pine forests
In the Southwest, few silvicultural treatments are implemented for the sole objective of reducing stand risk or susceptibility to bark beetles. Exceptions include Forest Health Protection (FHP)-funded projects (State and Private Forestry, USDA Forest Service) in

high value settings such as developed recreation (e.g., campgrounds) and administrative sites. The majority of federal funding for vegetation management is geared towards fuels reduction and forest health restoration projects.

Fuels reduction treatments in the wildland urban interface (WUI)

Most funding for vegetation management in southwestern ponderosa pine forests is expended on fuels reduction treatments, such as thinning from below, particularly in the WUI. While the primary objective of these treatments is to reduce the risk of catastrophic wildland fires and damage to homes and other structures (National Fire Plan 2004), these treatments are also often advocated as a strategy to reduce the susceptibility of individual trees and forest stands to bark beetle attack. However, there has not been a critical examination of how these treatments actually affect the short- and long-term susceptibility of stands to bark beetles. As thinning and prescribed fire prescriptions to reduce fuels can vary widely, there is reason to believe their effects on bark beetles will also vary. Thinning treatments with diameter caps of less than 41–46 cm can result in residual basal areas that are still in the moderate to high stand susceptibility for bark beetles that typically attack ponderosa pine. These treatments can also result in the creation of even aged stands comprised of large-diameter, mature trees that may be highly susceptible to bark beetle species such as western pine beetle, *D. brevicomis* LeConte, particularly during periods of extended drought. It is recommended that land managers, in cooperation with forest health professionals, monitor how bark beetles respond to such treatments in both the short- and long-term with the intent that silvicultural prescriptions can be developed that successfully achieve multiple goals with limited additional cost.

Forest health restoration treatments

Prescriptions for improving overall forest ecosystem health and function are also being implemented in southwestern ponderosa pine forests. In general, these treatments work to restore historic patterns of stand structure, fire intensity and fire frequency (Fulé et al. 2007). The resulting stand structure is typically patchier, clumpier and comprised of more uneven-aged stands compared with stand structures produced as a result of fuels reduction projects. Being that many of the stand hazard rating systems for ponderosa pine were developed in even-aged stands, there is a question as to how bark beetle activity might vary in response to these silvicultural systems (Negrón et al. 2008). Mountain pine beetle-caused tree mortality in uneven-aged ponderosa pine stands in the Black Hills of South Dakota and Wyoming was found to be positively correlated with basal area and ponderosa pine stand density index, which is similar to previous findings in even-aged stands (Schmid and Mata 2005). However, in contrast to even-aged stands where it is the total contribution of ponderosa pine that affects stand susceptibility, Negrón et al. (2008) concluded that densities (basal area) comprised of mid- to large-sized trees make a stand more susceptible to bark beetle attack in uneven-aged stands. Thus, akin to the recommendation for short- and long-term monitoring of bark beetle activity following fuels reduction treatments, additional case history studies of bark beetle responses to forest health restoration treatments seem prudent.

Research and Development

In a research context, bark beetle responses to vegetation management treatments must be considered at three spatial scales (i.e., individual tree, stand and landscape) and at least two temporal scales (i.e., short-term and long-term). Typically, research and development (R&D) efforts have concentrated on short-term (e.g., 1–5 years post-treatment) responses using small scale plots (e.g., ≤ 4 ha) indicative of stand level conditions. Given today's resource constraints, this is most appropriate, but not without certain limitations. For example, Schmid and Mata (2005) suggested results obtained from 1-ha plots within their Black Hills thinning study may be confounded by the fact that plots were surrounded by extensive areas of unmanaged forest where bark beetle populations were epidemic. They stated that reductions in long-term tree mortality will be accomplished when an area of sufficient size is managed so that thinned stands are separated from unmanaged stands by natural buffers or those of lower tree density. Several studies are being conducted at larger spatial scales (e.g., 10–100 ha) that represent more realistic management scenarios, but while data from such studies are highly desirable they come at significant cost.

Forest health specialists recognize long-term reductions in stand susceptibility to bark beetle attack achieved through vegetation management practices often occur at the cost of short-term increases in bark beetle-caused tree mortality. For example, as previously indicated, several bark beetle species are attracted to slash and/or host volatiles produced during thinning operations. While describing short-term bark beetle responses to vegetation management treatments are important, more important is the determination of long-term impacts on the amount and distribution of bark beetle-caused tree mortality as this influences fuel reduction targets, forest productivity and forest sustainability. One caveat is that long-term studies require long-term commitments in funding and staffing generally with relatively few accolades over time (i.e., presentations and publications) for the individual scientists and sponsoring agents involved. While the tremendous value of long-term studies is fully recognized, few funding sources are available for maintaining them.

In preparation for this presentation, we polled several of our colleagues in FHP to determine what they considered to be primary needs for research. Among vegetation management treatments, responses concentrated on the application of mechanical thinning and prescribed fire and their effects on the amount and distribution of bark beetle-caused tree mortality at three spatial scales (Table 1).

Table 1—Examples of research needs identified by Forest Health Protection, 2007

Research Question	Spatial Scale	Temporal Scale
What are the benefits of "individual tree culturing" to reduce the risk of western pine beetle attack on large diameter ponderosa pine in the Pacific Northwest?	Tree	Short and long-term
What is the probability of bark beetle attack on individual trees following prescribed fire? What can be done to limit any negative impacts?	Tree	Short and long-term
How does the application of prescribed fire influence the amount and distribution of bark beetle-caused tree mortality?	Stand	Short and long-term
What specific thinning treatments best meet long-term bark beetle management objectives?	Stand and landscape	Long-term
Are thinning treatments implemented during a bark beetle outbreak effective in the short- and/or long-term?	Stand and landscape	Short and long-term
How much of a landscape needs to be treated? Where will treatments be most effective?	Landscape	Long-term
Are there combinations of treatments that also satisfy other resource objectives?	Landscape	Long-term

The tools and methods by which thinning is implemented are quite diverse, and their application can result in significantly different stand structures and compositions. Depending on the insect species of concern, each method would have a functionally different response on the abundance and distribution of preferred hosts as well as that of the insect herbivore. For example, Whitehead and Russo (2005) suggested that increases in resin production and tree vigor following thinning were not as important in reducing mountain pine beetle-caused tree mortality in lodgepole pine stands as reductions in the number of initiated attacks, which is more likely associated with inter-tree spacing. In western North America, thinning has long been advocated as a preventive measure to alleviate or reduce the amount of bark beetle-caused tree mortality (Fettig et al. 2007).

Prescribed fire is often used to reduce the buildup of hazardous fuels, enhance wildlife habitat, improve grazing, thin overstocked stands, control some insects and diseases, prepare sites for regeneration and restore fire-adapted forest ecosystems. Forest managers must plan and execute prescribed burns carefully in order to minimize injury to desirable residual trees while still fulfilling management objectives. Bark beetles are often considered the most important mortality agent following prescribed fires, and mixed-severity wildfires, in coniferous forests (Parker et al. 2006). It has been our experience that gross generalizations concerning bark beetle responses to prescribed fire at the stand level are misleading as the bark beetle assemblages present within and adjacent to treated areas are of primary importance.

The research question "Are there combinations of treatments that also satisfy other resource objectives?" (Table 1) is particularly important and worthy of further discussion. In recent years, relatively few resources have been available to conduct thinnings specifically for bark beetle management (i.e., with consideration to residual tree distributions and densities within the context of lowering stand susceptibility to bark beetle attack). Therefore, it seems appropriate that forest health specialists should be working with fuel managers to determine if the application of SPLATs and SPOTs technology (i.e., Strategically Placed Landscape Area Treatments and Strategic Placement of Treatments as defined in fireshed assessments) used in fuels management could be adjusted to meet other forest health concerns. To our knowledge, this is not currently being done in the western US.

 We polled several of our colleagues in the Western Bark Beetle Research Group (WBBRG) to determine what studies were currently being conducted to identify bark beetle responses to vegetation management treatments (Table 2). It is encouraging that several studies will provide answers to questions posed in Table 1 and/or fill research gaps identified elsewhere (Fettig et al. 2007). For example, Massey and Wygant (1954) first reported the mean diameter of attacked Engelmann spruce, *Picea engelmannii* Parry ex Engelm., decreased during a spruce beetle, *D. rufipennis* (Kirby), outbreak thereby suggesting a preference by spruce beetle for larger diameter trees. Today, stands growing on well-drained sites and with a mean diameter at breast height (1.37 m) of live spruce > 25.4 cm being > 40.6 cm (i.e., large-diameter trees), basal areas > 34.3 m^2/ha and proportions of spruce > 65% are considered more susceptible to spruce beetle attack (Schmid and Frye 1976). However, no experiments have specifically been conducted to determine the effects of thinning on spruce beetle activity in Engelmann spruce stands. To generate such data within a completely randomized or randomized complete block design would take years or perhaps decades to establish the scientific infrastructure and await spruce beetle populations to challenge the experiment in a manner sufficient to determine differences in susceptibility among treatments. Alternatively, to address this knowledge gap Matt Hansen and Jose Negrón of WBBRG have recently initiated a retrospective study to determine the efficacy of silvicultural treatments in reducing stand-level spruce beetle-caused tree mortality, and to quantify post-outbreak stand characteristics among a variety of treatment types including unmanaged stands. Twenty-six pairs of previously treated and untreated plots have been installed in Arizona, Utah and Wyoming.

Table 2—Examples of ongoing research led by the Western Bark Beetle Research Group, 2007

Research Projects	Primary Invesigator(s)
Effects of silvicultural treatments on levels of spruce beetle-caused tree mortality in the Rocky Mountains	Hansen and Negrón
Tools for analyzing landscape-level fuels treatment scenarios and their effects on bark beetle-caused tree mortality	Hayes
Impacts of silvicultural treatments on defensive chemicals in stressed ponderosa and lodgepole pines and impacts on bark beetle host tree selection	Kelsey; Seybold
Factors associated with bark beetle-caused tree mortality at multiple spatial scales	Bentz; Fettig; Hansen; Negrón
Interactions among bark beetles and other disturbances to improve management approaches	Lundquist; Negrón; Seybold
Development of management guidelines to help reduce tree mortality due to bark beetle infestations after the application of prescribed fire	Bentz; Fettig; Hansen; Hayes; Kelsey; Lundquist; Negrón; Niwa
Thinning strategies for reducing the risk of bark beetle attack in Eastside pine and Sierra Nevada mixed conifer forests	Fettig

The "Ips-n-chips" Study

The *Ips*-n-chips study serves as a successful model for collaborative research between FHP and FS R&D (*see* Fettig et al. 2006). We share the genesis of this study as well as its results and impacts hoping that its serves as a model of success for similar studies conducted within the framework of WBBRG.

In recent years, unusually large and catastrophic wildfires have heightened public concern. Federal and state hazardous fuel reduction programs have increased accordingly to reduce the risk, extent and severity of these events, particularly in the WUI. Because sufficient markets have yet to be developed for small dimensional material in many locations, much of the tree biomass resulting from these treatments is not merchantable. In many areas, this material is cut and lopped (i.e., bole severed into short lengths and limbs removed) and/or chipped, and distributed on site. The amount of total biomass on the site may be unchanged, but the torching potential (i.e., the initiation of crown fire activity) and rate of potential crown fire spread is significantly reduced. However, these actions result in increased amounts of host material (slash) and host volatiles (from slash and chips) that may concentrate certain bark beetle species in these areas.

In early 2002, Joel McMillin and John Anhold (Forest Health Protection, USDA Forest Service, Flagstaff, AZ) were contacted regarding what appeared to be excessive amounts of bark beetle-caused tree mortality resulting from the chipping of unmerchantable trees during fuel reduction treatments in the WUI surrounding Flagstaff, Arizona. Through several site visits and a preliminary study, they provided anecdotal evidence that several bark beetle species appeared to be attracted to stands where logging residues had recently been chipped (McMillin and Anhold, unpublished data). In 2003, FHP (McMillin and Anhold) and the Pacific Southwest Research Station (Fettig) joined forces to examine the effects of several mechanical fuel reduction treatments on the activity of bark beetles in ponderosa pine forests located in Arizona and California. Treatments were applied in both late spring (April-May) and late summer (August-September) and included: (1) thinned biomass chipped and randomly dispersed within each 0.4 ha plot; (2) thinned biomass chipped, randomly dispersed within each plot and raked 2 m from the base of residual trees; (3) thinned biomass lopped-and-scattered (thinned trees cut into 1–2 m lengths) within each plot; and (4) an untreated control. The mean percentage of residual trees attacked by bark beetles ranged from 2.0% (untreated control) to 30.2% (plots thinned in spring with all biomass chipped). A three-fold increase in the percentage of trees attacked by bark beetles was observed in chipped versus lopped-and-scattered plots. Bark beetle colonization of residual trees was higher during spring treatments, which corresponded with peak adult beetle flight periods as measured by funnel trap captures. Raking chips away from the base of residual trees did not significantly affect attack rates. In a laboratory study, the quantities of β-pinene, 3-carene, α-pinene and myrcene eluting from chips greatly exceeded those from lopped-and-piled slash during each of 15 sample periods. These laboratory results may, in part, explain the bark beetle responses observed in chipping treatments as many of these monoterpenes are attractive, or enhance attraction in the presence of aggregation pheromone components, for several bark beetles.

Despite higher levels of bark beetle attack in chipped plots, no significant differences in tree mortality were observed among treatments during the first two years of this study. However, the authors commented that negative effects of prolonged and large numbers of red turpentine beetle, *D. valens* LeConte, attacks, among others, on individual tree health may not be realized for some time (Fettig et al. 2006), and continued monitoring these plots for bark beetle-caused tree mortality on an annual basis. During 2005 and 2006, a significant treatment effect was observed with significantly higher levels of bark beetle-caused tree mortality observed in plots chipped in spring than plots chipped in fall or those lopped-and-scattered in fall. Cumulatively (2003–2006), a significant treatment effect was also observed with significantly higher levels of bark beetle-caused tree mortality occurring in plots chipped in spring (6.1 ± 1.7 percent) than those lopped-and-scattered in fall (1.4 ± 0.8 percent).

Based on this study, guidelines were developed for minimizing tree losses due to bark beetle infestation following chipping (DeGomez et al. 2008). Again, we feel this study serves as a fruitful framework in which to conduct research within the context of WBBRG. We hope it serves as an example of one of many productive partnerships to come as a result of formation of the WBBRG.

Acknowledgements

We thank John Lundquist (Forest Health Protection and Pacific Northwest Research Station, USDA Forest Service, Anchorage, AK) for organizing and inviting us to participate in this symposium. We also thank our colleagues in FHP and WBBRG who responded to our inquiries for information in preparation for this presentation. John Anhold (Forest Health Protection, USDA Forest Service, Flagstaff, AZ), Chris Dabney (Pacific Southwest Research Station, USDA Forest Service, Placerville, CA), and John Nowak (Forest Health Protection, USDA Forest Service, Asheville, NC) provided critical reviews, which improved earlier versions of this manuscript. Much of the information discussed herein was synthesized from two of our recent publications (Fettig et al. 2006, 2007) and we wish to acknowledge the numerous contributions of our co-authors and cooperators to them. Funding sources for these publications and associated research efforts came from a USDA Forest Service Special Technology Development Program grant (R3-2003-01), the Pacific Southwest Research Station, Forest Health Protection and Washington Office, USDA Forest Service.

Literature Cited

Agee, J.K.; Skinner, C.N. 2005. Basic principles of forest fuel reduction treatments. Forest Ecology and Management. 211: 83–96.

Berryman, A.A. 1982. Population dynamics of bark beetles. In: Mitton, J.B.; Sturgeon, K.B., eds. Bark beetles in North American conifers. Austin, TX: University of Texas Press: 264–314.

Billings, R.F. 1995. Direct control of the southern pine beetle: rationale, effectiveness at the landscape level, and implications for future use of semiochemicals. In: Hain, F.P.; Salom, S.M.; Ravlin, W.F.; Payne, T.L.; Raffa, K.F., eds. Proceedings, IUFRO conference: behavior, population dynamics and control of forest insects, February 6–11, 1994, Maui, HI: 313–329.

Breece, C.R.; Kolb, T.E.; Dickson, B.G.; McMillin, J.D.; Clancy, K.M. 2008. Prescribed fire effects on bark beetle activity and tree mortality in southwestern ponderosa pine forests. Forest Ecology and Management. 255: 119–128.

Cocke, A.E.; Fulé, P.Z.; Crouse, J.E. 2005. Forest change on a steep mountain gradient after extended fire exclusion: San Francisco Peaks, Arizona, USA. Journal of Applied Ecology. 42: 814–823.

DeGomez, T.; Fettig, C.J.; McMillin, J.D.; Anhold, J.A.; Hayes, C. 2008. Managing slash to minimize colonization of residual leave trees by *Ips* and other bark beetle species following thinning in southwestern ponderosa pine. Tucson, AZ: University of Arizona, College of Agriculture and Life Sciences Bulletin. 21 p.

Eis, S.; Harman, E.H.; Ebell, L.F. 1965. Relation between cone production and diameter increment of Douglas-fir (*Pseudotsuga menziesii* [Mirb.] Franco), grand fir (*Abies grandis* [Dougl.] Lindl.), and western white pine (*Pinus monticola* Dougl.). Canadian Journal of Botany. 43: 1553–1559.

Fettig, C.J.; McMillin, J.D.; Anhold, J.A.; Hamud, S.M.; Borys, R.R.; Dabney, C.P.; Seybold, S.J. 2006. The effects of mechanical fuel reduction treatments on the activity of bark beetles (Coleoptera: Scolytidae) infesting ponderosa pine. Forest Ecology and Management. 230: 55-68.

Fettig, C.J.; Klepzig, K.D.; Billings, R.F.; Munson, A.S.; Nebeker, T.E.; Negrón, J.F.; Nowak, J.T. 2007. The effectiveness of vegetation management practices for prevention and control of bark beetle outbreaks in coniferous forests of the western and southern United States. Forest Ecology and Management. 238: 24–53.

Fogel, R.; Hunt, G. 1979. Fungal and arboreal biomass in a western Oregon Douglas-fir ecosystem: distribution patterns and turnover. Canadian Journal of Forest Research. 9: 245–256.

Franceschi, V.R.; Krokene, P.; Christiansen, E.; Krekling, T. 2005. Anatomical and chemical defenses of conifer bark against bark beetles and other pests. New Phytologist. 167: 353–376.

Fulé, P.Z.; Roccaforte, J.P.; Covington, W.W. 2007. Post-treatment tree mortality after forest ecological restoration, Arizona, United States. Environmental Management. 40: 623–634.

Furniss, R.L.; Carolin, V.M. 1977. Western forest insects. Misc. Pub. 1339. Washington, DC: U.S. Department of Agriculture, Forest Service. 654 p.

Kegley, S.J.; Livingston, R.L.; Gibson, K.E. 1997. Pine engraver, *Ips pini* (Say), in the United States. Forest Insect & Disease Leaflet 122. Missoula, MT: U.S. Department of Agriculture, Forest Service, Northern Region. 5 p.

Kramer, P.J.; Kozlowski, T.T. 1979. Physiology of woody plants. New York: Academic Press Inc. 811 p.

Livingston, R.L. 1979. The pine engraver *Ips pini* (Say) in Idaho, life history, habits, and management recommendations. Report 79-3. Coeur d'Alene, ID: Idaho Department of Lands. 7 p.

Massey, C.L.; Wygant, N.D. 1954. Biology and control of the Engelmann spruce beetle in Colorado. Circular 944. Washington, DC: U.S. Department of Agriculture, Forest Service. 35 p.

Mattson, W.J., Jr. 1977. The role of arthropods in forest ecosystems. New York: Springer-Verlag. 104 p.

Mattson, W.J., Jr.; Addy, N.D. 1975. Phytophagous insects as regulators of forest primary production. Science. 90: 515–522.

Mattson, W.J., Jr.; Niemela, P.; Rousi, M. 1996. Dynamics of forest herbivory: quest for pattern and principle. Gen. Tech. Rep. NC-GTR-183. St. Paul, MN: U.S. Department of Agriculture, Forest Service, North Central Research Station. 286 p.

McMillin, J.D.; Allen, K.K.; Long, D.F.; Harris, J.L.; Negrón, J.F. 2003. Effects of western balsam bark beetle on spruce-fir forests of north-central Wyoming. Western Journal of Applied Forestry. 18: 259-266.

Mitchell, R.G.; Waring, R.H.; Pitman, G.B. 1983. Thinning lodgepole pine increases the vigor and resistance to mountain pine beetle. Forest Science. 29: 204–211.

National Fire Plan. 2004. Hazardous Fuel Reduction: Fuel Reduction and Restoration Treatments. 2 p.

Negrón, J.F.; Allen, K.; Cook, B.; Withthrow, J.R., Jr. 2008. Susceptibility of ponderosa pine, *Pinus ponderosa* Dougl. ex Laws., to mountain pine beetle, *Dendroctonus ponderosae* Hopkins, attack in uneven-aged stands in the Black Hills of South Dakota and Wyoming USA. Forest Ecology and Management. 254: 327–334.

Oliver, C.D.; Larson, B.C. 1996. Forest stand dynamics. New York: John Wiley & Sons Inc. 520 p.

Parker, D.L. 1991. Integrated pest management guide: Arizona fivespined ips, *Ips lecontei* Swaine, and pine engraver, *Ips pini* (Say) in ponderosa pine. R3-91-8. Albuquerque, NM: U.S. Department of Agriculture, Forest Service, Southwestern Region. 17 p.

Parker, T.J.; Clancy, K.M.; Mathiasen, R.L. 2006. Interactions among fire, insects and pathogens in coniferous forests of the interior western United States and Canada. Agricultural and Forest Entomology. 8: 167–189.

Schmid, J.M.; Frye, R.H. 1976. Stand ratings for spruce beetle. Gen. Tech. Rep. RM-GTR-309. Fort Collins, CO: U.S. Department of Agriculture, Forest Service, Rocky Mountain Forest and Range Experiment Station. 4 p.

Schmid, J.M.; Mata, S.A. 2005. Mountain pine beetle-caused tree mortality in partially cut plots surrounded by unmanaged stands. Res. Pap. RM-54. Fort Collins, CO: U.S. Department of Agriculture, Forest Service, Rocky Mountain Research Station. 11 p.

Taylor, S.W., Carroll, A. 2004. Disturbance, forest age, and mountain pine beetle outbreak dynamics in BC: a historical perspective. In: Shore, T.L.; Brooks, J.E.; Stone, J.E., eds. Proceedings of the mountain pine beetle symposium: challenges and solutions, Kelowna, BC, 30–31 October 2003. Information Report No. BC-X-399. Victoria, BC: Natural Resources Canada, Canadian Forest Service, Pacific Forestry Centre. 11 p.

USDA Forest Service. 2005. Forest insect and disease conditions in the United States 2004. Washington, DC: U.S. Department of Agriculture, Forest Service, Forest Health Protection. 142 p.

Wallin, K.F.; Raffa, K.F. 2000. Influence of host chemicals and internal physiology on the multiple steps of postlanding host acceptance behavior of *Ips pini* (Coleoptera: Scolytidae). Environmental Entomology. 29: 442–453.

Waring, R.H.; Schlesinger, W.H. 1985. Forest ecosystems: concepts and management. New York: Academic Press. 340 p.

Whitehead, R.J.; Russo, G.L. 2005. "Beetle-proofed" lodgepole pine stands in interior British Columbia have less damage from mountain pine beetle. Report BC-X-402. Victoria, BC: Natural Resources Canada, Canadian Forest Service, Pacific Forestry Centre. 17 p.

Wood, S.L. 1982. The bark and ambrosia beetles of North and Central America (Coleoptera: Scolytidae), a taxonomic monograph. Provo, UT: Great Basin Naturalist Memoirs 6. 1359 p.

Bark Beetles in a Changing Climate[1]

John E. Lundquist and Barbara J. Bentz[2]

Abstract

Over the past decade, native bark beetles (Coleoptera: Curculionidae) have killed billions of trees across millions of hectares of forest from Alaska to Mexico. Although bark beetle infestations are a regular force of natural change in forested ecosystems, several current outbreaks occurring simultaneously across western North America are the largest and most severe in recorded history. Bark beetle ecology is complex and dynamic, and a variety of circumstances must coincide for a large scale bark beetle outbreak. While outbreak dynamics vary from bark beetle species to bark beetle species and from forest type to forest type, a combination of several factors appear to be driving current outbreaks, including a changing climate.

Keywords: Global warming, climate change, latitudinal gradient, climate models, climate normals.

[1] The genesis of this manuscript was a presentation by the authors at the Western Bark Beetle Research Group—A Unique Collaboration with Forest Health Protection Symposium, Society of American Foresters Conference, 23–28 October 2007, Portland, OR.

[2] **John E. Lundquist** is an Entomologist, USDA Forest Service, R-10 Forest Health Protection and Pacific Northwest Research Station, Anchorage, AK; email: jlundquist@fs.fed.us. **Barbara J. Bentz** is a Research Entomologist, USDA Forest Service, Rocky Mountain Research Station, Logan, UT; email: bbentz@fs.fed.us.

Introduction

Over the past 100 years, global average temperature has risen by 0.74°C (0.56–0.92°C range). The greatest increase has occurred during the last two decades and experts say increases will continue (CIRMOUNT 2006). Predictions for increasing average global temperatures range from 1.0 °C to 4 °C over the next 100 years (Houghton et al. 2001).

There has also been a dramatic increase in the number of publications on climate change. An internet search of "climate change" finds over 60,000 hits in 2007 alone! Climate change may be one of the most focused topic areas in living history.

The amazing interest in this topic is not easy to explain. Most would agree that the biological understanding and science underlying the climate change phenomenon has existed for at least a couple of decades (Houghton et al. 2001). But science alone has been inadequate in evoking such a response. Politics and the media apparently lined up just right with science causing climate change to emerge from "science" to a truly popular phenomenon (Boykoff 2007).

What is climate change?

"Climate" refers to the average state of the weather. Common weather phenomena are temperature, rain, snow, fog, wind, cloud, dust storms, and events such as tornadoes, hurricanes and ice storms. Weather usually refers to activity of these phenomena over short periods of time (hours or days) in localized areas. Climate refers to average atmospheric conditions over longer periods of time and involves broad areas. Although climate change is usually portrayed as increasing temperature, it actually expresses itself in many other ways as well; e.g., as changes in precipitation, UV-B radiation, atmospheric CO_2, nitrogen deposition, and others.

We see effects of climate change easier than we can experience a changing climate. These changes impact many things (Kolbert 2006, Parmesan 2006, Roy and Sparks 2000, Wohlforth 2002). We have heard about glaciers melting, flowers blooming earlier than in previous years, ocean levels rising and ocean-front villages washing to the sea, butterfly distributions migrating north (and one or two going south), and bark beetle outbreaks killing millions of hectares of forests. Climate is one of those unique phenomena that has the ability to effect nearly everything. It has been referred to as the "ultimate integrative field".

Why should we be interested in the effects of climate on herbivorous insects, like bark beetles?

Insects have short life cycles, resulting in dozens or hundreds of generations in the time it takes most higher plants to complete one generation. Because insect life cycles are environmentally driven, a change in climate can significantly influence insect population timing and density. Monitoring insect population trends can then be used as an indirect measure of climate change. Furthermore, insects occur nearly everywhere and many can be studied year round. Insect pests can shape or change ecosystem structure and function, and, in doing so, act as catalysts of change. Insects can help maintain or sustain ecosystems; displace or remove components of ecosystems; or lead to replacement of existing ecosystems. Insects can respond to long-term subtle shifts in their environment that are commonly so subtle that they cannot be directly experienced by humans. In short, insects can serve as a convenient bioindicator of climate change.

Bark beetles, in particular, create the most visible of insect disturbances in a forest because they kill trees, lots of trees, and their impacts vary across all ecosystem services provided by a forest (Fettig et al. 2007). They influence nutrient cycling, energy flow, decomposition and other supporting services of ecosystems. They affect wood production and other provisioning services and goods of ecosystems. They influence water production, snow distribution and other regulating functions of ecosystems. They impact recreational experiences and other cultural services. They react quickly to changes in climate; much faster than higher plants including trees. Many bark beetle species have geographic distributions less extensive than their tree hosts (Ayres and Lombardero 2000), which suggests distributions could rapidly shift with climate change (Carroll et al. 2004).

Climate change can directly affect bark beetle phenology and winter mortality, resulting in shifts in length and number of annual life cycles. Bark beetle communities will also be affected including predator/prey relationships, interactions with symbiotic fungi, forest structure, and forest vigor. Changes in temperature, precipitation and atmospheric gases will undoubtedly affect host tree defenses as well, possibly resulting in changes to bark beetle host specificity and geographic distribution. Rapid changes in climate may also result in genetic adaptations that create metapopulations (Balanya et al. 2006, Bradshaw and Holzapfel 2006). Spatial and temporal synchrony of beetles and their host trees may also be disrupted.

Dramatic increase in outbreaks of bark beetles in the West
Western U.S. states and Canadian provinces have recently seen a significant increase in bark beetle activity. Examples include pinyon ips (*Ips confusus* LeConte) on pinyon pine (mostly *Pinus monophylla* Torr. & Frem. and *P. edulis* Engelm.) in the southwestern U.S. (Breshears et al. 2005), spruce beetle (*Dendroctonus rufipennis* Kirby) in Alaska (Werner et al. 2006), mountain pine beetle (*D. ponderosae* Hopkins) along the Rocky Mountain Front Range in Colorado (Negrón and Popp 2004), in high elevation forests (Gibson 2006, Bentz and Schen-Langenheim 2007), and in interior British Columbia (Westfall and Ebata 2008). Climate change has been implicated as a major influencing factor (Berg et al. 2006, Breshears et al. 2005, Nijhuis 2004). Proving a direct correlation between climate and bark beetle outbreaks, however, is a difficult task.

Understanding how climate affects the mechanics of bark beetle outbreaks is a challenge
Climate affects everything in an already complicated biological system. For a bark beetle outbreak to occur, there must be suitable climate for several years, an active beetle population, and an extensive area of host trees of appropriate age, size and species (Fettig et al. 2007). Temperature, moisture and other climatic elements

symbolic of a changing climate can affect these requirements for a bark beetle outbreak. Outbreaks are often non-linear, unpredictable, sometimes unexpected events (Logan et al. 2003). Bark beetle outbreaks result from a unique combination of conditions at a variety of scales (Raffa et al. 2008).

Elevated temperature and shifting precipitation patterns, in particular, appear to be influencing recent and current bark beetle outbreaks (Régnière and Bentz 2007, Shaw et al. 2005). Elevated temperatures can speed up reproductive and growth cycles and reduce cold-induced mortality during cold snaps (Bentz and Mullins 1999, Bentz et al. 1991, Logan and Bentz 1999). Although the relationship is nonlinear, prolonged drought can weaken trees, making them more susceptible to bark beetle attacks (Breshears et al. 2005, Mattson and Haack 1987, Waring and Cobb 1992).

Because bark beetles are one component of a rich community comprising forest ecosystems, to fully understand climate change effects on bark beetles and hence forest ecosystems, we need to consider how climate change influences biotic interactions of symbiosis, competition, predation and other dynamic disturbance processes (Botkin et al. 2007). For example, Six and Bentz (2007) observed that temperature determines the relative presence of symbiotic fungi associated with mountain pine beetle. Although relationships are unclear at this time, it is obvious that climate change effects on fungal populations will have a cascading effect on mountain pine beetle population success. Effects of climate change on other critical components of bark beetle communities, including predators and parasites, are also unclear.

Predicting climate and weather events of the future is a very difficult task
Forecasts are less reliable the further out in time they project. Small, seemingly insignificant, changes can amplify to become major system shifts, which are unpredictable (Burkett et al. 2005). Because of a sensitivity to small changes, it will never be possible to make perfect forecasts (Holling 2001), although there still is much potential for improvement. Useful predictions of future insect activity will depend on reliable predictions of weather and a good understanding of cause/effect relations between weather/climate and insect physiology, behavior, and ecology (Stireman et al. 2005). Predicting the future involves some very complex mathematical models and very advanced, high capacity computers (McKenney et al. 2003, Rehfeldt et al. 2006, Williams and Liebhold 2002).

Mechanistic mathematical models have been developed to describe and predict mountain pine beetle phenology (Bentz et al. 1991, Gilbert et al. 2004, Jenkins et al. 2001, Logan and Bentz 1999, Powell et al. 2000, Safranyik et al. 1975) and cold tolerance (Régnière and Bentz 2007), and spruce beetle voltinism (Hansen et al. 2001b). These models have been implemented within the BioSim (Régnière and St-Amant 2007) modeling framework, enabling landscape-scale projections of population success given daily temperatures for the duration of a generation, one, two or three years depending on bark beetle species and geographic location.

Model results using climate-changed normals suggest that the probability of mountain pine beetle temperature-dependent survival in western U. S. over the next 25 yrs will generally increase. High-elevation forests will experience the greatest increase in probability of mountain pine beetle survival. The biggest increase in univoltine spruce beetle populations, and thus exponential population growth, is predicted to occur in Alaska and high-elevation areas of the western U.S. Historically, spruce beetle has had a two-year, or in some cases three-year, life cycle in these areas.

Our ability to predict western U.S. bark beetle response to climate change is limited by a lack of data on species-specific temperature-dependent developmental processes. As described above, we do have models for mountain pine beetle and spruce beetle, although additional research is needed to parameterize existing models to account for regional genetic differences in population response to temperature. For other bark beetle species, our current ability to forecast climate change effects on population dynamics is almost entirely qualitative.

What can we expect to happen to bark beetle populations as climate changes?
Many possible scenarios have been proposed. In general, insect outbreaks are probably going to increase in number and severity (Ayres and Lombardero 2000, Stireman et al. 2005). Interactions between insects and their natural control agents (parasites, pathogens, predators, parasitoids) may be disrupted resulting in positive or negative effects on insect populations (Malmstrom and Raffa 2000). Host plant and insect phenological synchrony may be disrupted. Winter survival of insects may increase (Regniere and Bentz 2007, Williams and Liebhold 2002). Observed genetic adaptation to local environmental conditions (Bentz et al. 2001) suggest that bark beetles could rapidly respond to a changing climate. Exotic insects will have more opportunities to invade new areas and previously innocuous insects may shift hosts and become pests (Pernek et al. 2008). Distributions will shift northward in latitude and upward in elevation. Bark beetle species currently restricted to the southern U.S. and Mexico could expand northward. Northernmost forests will be affected first and most severely (Thomas et al. 2006).

Bark beetle management under a changing climate
Managers want to know what can be done to hedge against future effects under such a cloud of uncertainty. Several management coping strategies have been proposed including a change in forest structure and age patterns across landscapes, altered species composition and diversity, reduction in invasive species populations, prompt action when new invasives are detected, planting late successional species, and many others (Spittlehouse and Stewart 2003). Most of these suggestions are based on logic alone, since unprecedented conditions are facing managers. Few are based on statistically rigorous experimental research. Managers must be willing to accept that climate change will result in novel environmental conditions never experienced by current forest ecosystems, and dynamic strategies that enhance ecosystem adaptability will be required (Millar et al. 2007).

Conclusions

Management traditionally has been aimed at recreating the past using such concepts as historical range of variability (Choi 2007). We treat the past as a stable state. We are beginning to realize now that we have no such stable state under a changing climate. We are tremendously challenged to predict what future suitably resilient environments will look like. The future is a moving target. One thing that is highly probable... the climate will change. Which way it changes is a question on many researchers and practitioners minds. Extrapolating the climate versus time curve is a challenging effort, and some believe that taking actions in response to climate change can create a bigger risk than doing nothing (Spittlehouse and Stewart 2003).

There are many unanswered questions about potential effects of climate change on western bark beetle populations. Many will be difficult, perhaps impossible, to answer. Management of western U.S. forest ecosystems should be based on the best available science, a prospect facilitated by scientists within U.S. Forest Service Research and Development Western Bark Beetle Research Group.

Acknowledgements

We thank Ken Gibson, Mark Schultz, Chris Fettig, Aileen Holthaus, and Steve Patterson for reviewing earlier versions of this manuscript. This paper was originally presented at the 2007 Annual Meeting of the Society of American Foresters. We are grateful to SAF.

Literature Cited

Ayres, M.P.; Lombardero, M.J. 2000. Assessing the consequences of global change for forest disturbance from herbivores and pathogens. The Science of the Total Environment. 262: 263–286.

Balanya, J.; Oller, J.M.; Huey, R.B.; Gilchrist, G.W.; Serra, L. 2006. Global genetic tracks global climate warming in *Drosophila subobscura*. Science. 313: 1773–1775.

Bentz, B.J.; Mullins, D.E. 1999. Ecology of mountain pine beetle cold hardening in the Intermountain West. Environmental Entomology. 28(4): 577–587.

Bentz, B.J.; Schen-Langenheim, G. 2007. The mountain pine beetle and whitebark pine waltz: has the music changed? Proceedings of the Conference Whitebark Pine: A Pacific Coast Perspective. http://www.fs.fed.us/r6/nr/fid/wbpine/papers/2007-wbp-impacts-bentz.pdf.

Bentz, B.J.; Logan, J.A.; Amman, G.D. 1991. Temperature-dependent development of the mountain pine beetle (Coleoptera: Scolytidae) and simulation of its phenology. Canadian Entomologist. 123: 1083–1094.

Berg, E.E.; J.D. Henry; C.L.Fastie; A.D. De Volder; S.M. Matsuoka. 2006. Spruce beetle outbreaks on the Kenai Peninsula, Alaska, and Kluane National Park and Reserve, Yukon Territory: relationship to summer temperatures and regional differences in disturbance regimes. Forest Ecology and Management. 227: 219–232.

Botkin, D.B.; Saxe, H.; Araújo, M.B.; Betts, R.; Bradshaw, R.H.W.; Cedhagen, T.; Chesson, P.; Dawson, T.P.; Etterson, J.R.; Faith, D.P.; Ferrier, S.; Guisan, A.; Hansen, A. S.; Hilbert, D.W.; Loehle, C.; Margules, C.; New, M.; Sobel, M.J.; Stockwell, D.R.B. 2007. Forecasting the effects of global warming on biodiversity. BioScience. 57: 227–236

Boykoff, M. 2007. Changing patterns in climate change reporting in United States and United Kingdom media. Presentation at Carbonundrums: making sense of climate change reporting around the world. Reuters Institute for the Study of Journalism & Environmental Change. June 27, 2007. http://www.eci.ox.ac.uk/news/events/070727-carbonundrum/boykoff.pdf. (12 January 2009).

Bradshaw, W.E.; Holzapfel, C.M. 2006. Evolutionary response to rapid climate change. Science. 312: 1477–1478.

Breshears, D.D.; Cobb, N.S.; Rich, P.M.; Price, K.P.; Allen, C.D.; Balice, R.G.; Romme, W.H.; Kastens, J.H.; Floyd, M.L.; Belnap, J.; Anderson, J.J.; Myers, O.B.; Meyer, C.W. 2005. Regional vegetation die-off in response to global-change-type drought. Proceedings of the National Academy of Science. 102: 15144–15148.

Burkett, V.R.; Wilcox, D.A.; Stottlemyer, R.; Barrow, W.; Fagre, D.; Baron, J.; Price, J.; Nieldsen, J.L.; Allen, C.D.; Peterson, D.L.; Ruggerone, G.; Doyle, T. 2005. Nonlinear dynamics in ecosystem response to climatic change: case studies and policy implications. Ecological Complexity. 2: 357–394.

Carroll, A.L.; Taylor, S.W.; Regniere, J.; Safranyik, L. 2004. Effects of climate change on range expansion by the mountain pine beetle in British Columbia. In: Shore, T.L.; Brooks, J.E.; Stone, J.E., eds. Mountain pine beetle symposium: challenges and solutions. October 30–31, 2003, Kelowna, British Columbia. Information Report BC-X-399. Victoria, BC: Natural Resources Canada, Canadian Forest Service, Pacific Forestry Centre. 298 p.

Choi, Y.D. 2007. Restoration ecology to the future: a call for new paradigm. Restoration Ecology. 15(2): 351–353.

CIRMOUNT Committee. 2006. Mapping new terrain: climate change and America's West. Report of the Consortium for Integrated Climate Research in Western Mountains (CIRMOUNT). Misc. Pub., PSW-MISC-77. Albany, CA: US Department of Agriculture, Forest Service, Pacific Southwest Research Station. 29 p.

Fettig, C.J.; Klepzig, K.D.; Billings, R.F.; Munson, A.S.; Nebeker, T.E.; Negrón, J.F.; Nowak, J.T. 2007. The effectiveness of vegetation management practices for prevention and control of bark beetle outbreaks in coniferous forests of the western and southern United States. Forest Ecology and Management. 238: 24–53.

Gibson, K. 2006. Mountain pine beetle conditions in whitebark pine stands in the greater Yellowstone ecosystem, 2006. Numbered Report 06-03. Missoula, MT: U.S. Department of Agriculture, Forest Service, Forest Health Protection. 7 p.

Gilbert, E.; Powell, J.A.; Logan, J.A.; Bentz, B.J. 2004. Comparison of three models predicting developmental milestones given environmental and individual variation. Bulletin of Mathematical Biology. 66: 1821–1850.

Hansen, E.M.; Bentz, B.J.; Turner, D.L. 2001a. Physiological basis for flexible voltinism in the spruce beetle (Coleoptera: Scolytidae). Canadian Entomologist. 133: 805–817.

Hansen, E.M.; Bentz, B.J.; Turner, D.L. 2001b. Temperature-based model for predicting univoltine brood proportions in spruce beetle (Coleoptera: Scolytidae). Canadian Entomologist. 133: 827–841.

Holling, C.S. 2001. Understanding the complexity of economic, ecological, and social systems. Ecosystems. 4: 390–403.

Houghton, J.T.; Ding, Y.; Griggs, D.J.; Noguer, M.; van der Linden, P.J.; Dai, W.; Maskell, K.; Johnson, C.A., eds. 2001. Climate change 2001: the scientific basis. Contribution of Working Group I to the Third Assessment Report of the Intergovernmental Panel on Climate Change. Cambridge University Press, Cambridge and New York. 881 p.

Jenkins, J.L.; Powell, J.A.; Logan, J.A.; Bentz, B.J. 2001. Low seasonal temperatures promote life cycle synchronization. Bulletin of Mathematical Biology. 63: 573-595.

Kolbert, E. 2006. Field Notes from a Catastrophe: Man, Nature, and Climate Change. Atlantic Monthly Press, N.Y. 357 p.

Logan, J.A.; Regniere, J.; Powell, J.A. 2003. Assessing the impact of global warming on forest pest dynamics. Frontiers in Ecology and the Environment. 1(3): 130–137.

Logan, J.A.; Bentz, B.J. 1999. Model analysis of mountain pine beetle (Coleoptera: Scolytidae) seasonality. Environmental Entomology. 28(6): 924–934.

Malmstrom, C.M.; Raffa, K. F. 2000. Biotic disturbance agents in the boreal forest: considerations for vegetation change models. Global Change Biology. 6(1): 35–48.

Mattson, W.J.; Haack, R.A. 1987. The role of drought in outbreaks of plant-eating insects. BioScience. 37: 110–118.

McKenney, D.W.; Hopkin, A.A.; Campbell, K.L.; Mackey, B.G.; Roottit, R. 2003. Opportunities for improved risk assessments of exotic species in Canada using bioclimatic modeling. Environmental Monitoring and Assessment. 88: 445–461.

Millar, C.I.; Stephenson, N.L.; Stephens, S.L. 2007. Climate change and forests of the future: managing in the face of uncertainty. Ecological Applications. 17(8): 2145–2151.

Negrón, J.F.; Popp, J.B. 2004. Probability of ponderosa pine infestation by mountain pine beetle in the Colorado Front Range. Forest Ecology and Management. 191: 17–27.

Nijhuis, M. 2004. Global warming's unlikely harbingers. High Country News, July 19, 2004.

Parmesan C. 2006. Ecological and evolutionary responses to recent climate change. Annual Review of Ecology, Evolution and Systematics. 37: 637–669.

Pernek, M.; Pilas, I.; Vrbek, B.; Benko, M.; Hrasovec, B.; Milkovic, J. 2008. Forecasting the impact of the Gypsy moth on lowland hardwood forests by analyzing the cyclical pattern of population and climate data series. Forest Ecology and Management. 255: 1740-1748.

Powell, J.A.; Jenkins, J.L.; Logan, J.A.; Bentz, B.J. 2000. Seasonal temperature alone can synchronize life cycles. Bulletin of Mathematical Biology. 62: 977–998.

Raffa, K.F; Aukema, B.H.; Bentz, B.J.; Carroll, A.L.; Hicke, J.A.; Turner, M.G.; Romme, W.H. 2008. Cross-scale drivers of natural disturbances prone to anthropogenic amplification: dynamics of biome-wide bark beetle eruptions. BioScience. 58(6): 501–518.

Régnière, J.; Bentz, B.J. 2007. Modeling cold tolerance in the mountain pine beetle, *Dendroctonus ponderosae*. Journal of Insect Physiology. 53: 559–572.

Régnière J.; St-Amant, R. 2007. Stochastic simulation of daily air temperature and precipitation from monthly normals in North America north of Mexico. International Journal of Biometeorology. 51: 415–430.

Rehfeldt, G.E.; Crookston, N.L.; Warwell, M.V.; Evans, J.S. 2006. Empirical analyses of plant-climate relationships for the western United States. International Journal of Plant Science. 167: 1123–1150.

Roy, D.B; Sparks, T.H. 2000. Phenology of British butterflies and climate change. Global Change Biology. 6: 407–416.

Safranyik L.; Shrimpton D.M.; Whitney H.S. 1975. .An interpretation of the interactions between lodgepole pine, the mountain pine beetle, and its associated blue stain fungi in western Canada. In: Baumgartner D.M., ed. Management of lodgepole pine ecosystems. Pullman, WA: Washington State University Cooperative Extension Service: 406–428.

Shaw, J.D.; Steed, B.E.; DeBlander, L.T. 2005. Forest inventory and analysis (FIA) annual inventory answers the question: what is happening to pinyon-juniper woodlands? Journal of Forestry. 103: 280–285.

Six, D.L.; Bentz, B.J. 2007. Temperature determines the relative abundance of symbionts in a multipartite bark beetle-fungus symbiosis. Microbial Ecology. 54: 112–118.

Spittlehouse, D.L.; Stewart, R.B. 2003. Adaptation to climate change in forest management. BC Journal of Ecosystems and Management. 4(1). http://www.forrex.org/jem/2003/vol4/no1/art1.pdf.

Stireman, J.O. III; Dyer, L.A.; Janzen, D.H.; Singer, M.S.; Lill, J.T.; Marquis, R.J.; Ricklefs, R.E.; Gentry, G.L.; Hallwachs, W.; Coley, P.D.; Barone, J.A.; Greeney, H.F.; Connahs, H.; Barbosa, P.; Morais, H.C.; Diniz, I.R. 2005. Climatic unpredictability and parasitism of caterpillars: implications of global warming. Proceedings of the National Academy of Sciences. 102: 17384–17387.

Thomas, C.D.; Franco, A.M.A.; Hill, J.K. 2006. Range retractions and extinction in the face of climate warming. Trends in Ecology and Evolution. 21(8): 415–416.

Waring, G.L.; Cobb, N.S. 1992. The impact of plant stress on herbivore population dynamics. In: Bernays, E., ed. Insect-plant interactions, vol. IV. Boca Raton, FL: CRC Press: 167–226.

Werner, R.A.; Holsten, E.H. 1985. Effect of phloem temperature on development of spruce beetles in Alaska. In: Safrankik, L., ed. The role of the host in the population dynamics of forest insects. Victoria, BC: Forest Canada, Pacific Forest Centre: 155–163.

Werner, R.A.; Holsten, E.H.; Matsuoka, S.M.; Burnside, R.E. 2006. Spruce beetles and forest ecosystems in south-central Alaska: a review of 30 years of research. Forest Ecology and Management. 227: 195–206.

Westfall, J.; Ebata, T. 2008. Summary of forest health conditions in British Columbia - 2007. Pest Management Report No. 15. Victoria, BC: British Columbia Ministry of Forests, Forest Practices Branch. 81 p. http://www.llbc.leg.bc.ca/public/PubDocs/ bcdocs/354419/bc_forests_hlth_con_2007.pdf.

Williams, D.W.; Liebhold, A.M. 2002. Climate change and the outbreak ranges of two North American bark beetles. Agricultural and Forest Entomology. 4: 87–99.

Wohlforth, C. 2002. Spruce bark beetles and climate change. Alaska Magazine. March 2002. http://www.wohlforth.net/FullTextArticles.html (6 January 2009)

Fire and Bark Beetle Interactions[1]

Ken Gibson and José F. Negrón[2]

Abstract

Bark beetle populations are at outbreak conditions in many parts of the western United States and causing extensive tree mortality. Bark beetles interact with other disturbance agents in forest ecosystems, one of the primary being fires. In order to implement appropriate post-fire management of fire-damaged ecosystems, we need a better understanding of relationships between bark beetles and wildfire. Interactions can be one of two primary types: Fires can influence bark beetle populations directly by providing significant amounts of susceptible trees which may precipitate serious outbreaks; and effects of bark beetle outbreaks may influence likelihood and behavior of future fires. We examine various aspects of these interactions.

Keywords: Bark beetles, fire, fire-insect interactions.

[1] The genesis of this manuscript was a presentation by the authors at the Western Bark Beetle Research Group—A Unique Collaboration with Forest Health Protection Symposium, Society of American Foresters Conference, 23-28 October 2007, Portland, OR.

[2] **Ken Gibson** is an Entomologist with the USDA Forest Service, Forest Health Protection, P.O. Box 7669, Missoula, MT 59807; email: kgibson@fs.fed.us. **José F. Negrón** is a Research Entomologist with the USDA Forest Service, Rocky Mountain Research Station, 240 West Prospect Rd., Fort Collins, CO 80525; email: jnegron@fs.fed.us.

Introduction

The collective wildfire seasons over past decade have been some of the most widespread and damaging in recorded history. As such, wildfires unquestionably have had both short- and long-term effects on management activities in forested stands of the intermountain West. Some of those effects may be initiation of bark beetle outbreaks. In other cases, existing outbreaks may be prolonged. Land managers need to determine, to the extent possible, which trees are likely to succumb to fire damage, which might survive fire effects but be killed by bark beetles, and which others may survive them both. The sooner those assessments can be made and preventive or corrective measures implemented, the more successfully adverse effects will be avoided (Missoula Field Office 2000). The relationship between bark beetle-caused mortality and resultant effects on fire behavior continue to generate questions. These relationships will also be discussed.

Other authors in these proceedings have discussed current bark beetle conditions, in western coniferous forests, where extreme tree mortality occasionally occurs due to elevated insect populations (*see* Cain and Hayes 2008). If we want to develop and implement appropriate post-fire management of fire-damaged forest ecosystems, we will need a better understanding of relationships between bark beetles and wildfire. This interaction can take two primary forms: Fires can have a significant impact on population dynamics of bark beetles which in turn can cause tree injury; and occurrence of bark beetles have many effects in coniferous forest ecosystems—one of which may be influencing the likelihood and behavior of future fires through changes in stand structure, transformation of live fuels into dead fuels, and fuel arrangements. In this paper, we examine two commonly held assumptions—fires have a significant impact on population dynamics of beetles; and that bark beetle-caused mortality, likewise, has a significant impact on wildfire behavior.

Post-fire tree survivability and bark beetle interactions

Recently obtained research results can make prognoses of tree survival and appropriate management responses to both fire and threats from bark beetles more effective. Ryan (1982, 1989) has shown that the probability of tree survival is related to damage to crown, stem, or roots. Furthermore, amount of damage individual trees can sustain and still survive is dependent upon characteristics of its species (needle length and bark thickness), its size (diameter and height), and site factors on which it is growing. Research by Ryan, Harrington and Reinhardt has provided helpful means of predicting post-fire mortality based on species-specific characteristics (Harrington 1996, Reinhardt and Ryan 1989, Ryan and Amman 1994). Studies recently completed by Hood et al. (2007) and Sieg et al. (2006) have greatly helped answer survivability questions for two coniferous tree species—Douglas-fir and ponderosa pine, respectively.

In some cases, effects of earlier fires and management responses to bark beetle-induced mortality have served as valuable information sources. Included here are summaries of pertinent research results, useful historic precedents, and projects involving management activities implemented during previous post-fire evaluations. We have learned that recommendations must be general enough to have widespread applicability, yet specific enough to be locally worthwhile. Still, recommendations are subject to site-specific conditions that are often difficult to predict: fire effects on bark beetle hosts, weather one or two years post-fire, extant populations of host-specific bark beetle species, and interactions between all three.

Within the past decade, forested stands in the West, of all ownerships, have been both extensively and intensely affected. Fire damage, of varying severity, has extended to several million acres in each of the past ten years. Yearly, fires rage in some parts of the West from April through November. Even as fires burn, post-fire planning to deal with their aftermath must proceed. There is, and will continue to be, a need to address wildfire effects in forested stands, and perhaps even more critically, in the more-populated wildland-urban interface. What short- and long-term management decisions will be implemented and how; and how bark beetles will interact with fire-damaged trees are questions that must be answered—and the sooner the better.

Bark Beetle Considerations

Following wildfires, land managers are naturally concerned about tree survivability. We have also learned, in some situations, there is a high likelihood of bark beetles infesting fire-weakened trees (Parker et al. 2006). Bark beetle outbreaks following wildfires are not unprecedented, but neither are they certain. Several conditions must exist for bark beetles to take advantage of fire-damaged hosts:

1. There must be a sufficient supply of undamaged inner bark in fire-affected trees. If beetles' food supply, the bark and inner bark (phloem), becomes dry or scorched—often the case in stand-replacing fires or in thin-barked tree species—beetles will neither feed nor lay eggs in it.
2. Fires must occur at a time when beetles either are, or soon will be, in the adult stage and capable of infesting susceptible trees. Fires in late summer or early fall may occur after beetles have flown or may be colonized by wood borers and may therefore not be as suitable to bark beetles the following year. A recently killed tree's inner bark remains usable to beetles for a relatively short time. If not attacked while still "green," phloem may become too dry or otherwise unusable before the next flight season.
3. There must be a population of beetles within a reasonable distance to take advantage of weakened trees which become available.
4. Post-fire weather must be conducive to beetle survival and propagation.

Fire Survivability Case Studies

Because several conditions must be met for outbreak development, beetle epidemics following wildfires are not a foregone conclusion; but a few such outbreaks are well-documented. Douglas-fir beetle (*Dendroctonus pseudotsugae* Hopkins), spruce beetle (*D. rufipenis* (Kirby)), and pine engraver beetle (*Ips* spp.) outbreaks following wildfires in 1988, 1994, and 2000 became extensive and quite damaging in parts of Yellowstone National Park and Montana (Amman and Ryan 1991; Rasmussen et al. 1996; Ryan and Amman 1996; FHP, Northern Region, unpublished office reports).

Following 1988 Yellowstone National Park fires, Amman and Ryan (1991) concluded "The 1988 fires in the Greater Yellowstone Area killed many trees outright. Many more were subjected to sublethal injuries resulting in increased susceptibility to insect attack. Still other trees escaped fire injury but are exposed to the spread of insect attack from nearby injured trees." Rasmussen et al. (1996) showed "that bark beetle and delayed tree mortality due to fire injury significantly alter mosaics of green and fire-injured trees, that insect infestation increases with the percent of basal circumference killed by fire, and that bark beetle populations appear to increase in fire-injured trees and then infest uninjured trees."

Ryan and Reinhardt (1988) demonstrated that post-fire mortality can be predicted as a function of crown scorch and bark thickness for most western conifers and that probability of mortality increased with percentage of crown killed and decreased as bark thickness increased. Weatherby et al. (1994) used those relationships in an effort to evaluate tree survivability following the 1989 Lowman, ID fire. They found 82% of the ponderosa pine and 52% of the Douglas-fir survived the fire; but a significant portion was killed by bark beetles as opposed to direct fire effects.

Observations made following wildfires in western Montana have shown that Douglas-fir is likely to be killed by Douglas-fir beetles if cambium has been killed on half or more of the bole circumference. Occasionally, that damage may occur on large, lateral roots at or below the duff (Hood et al. 2007). Amman and Ryan (1991) showed that 71% of the Douglas-fir on their Yellowstone plots died—over twice as many as predicted by the model using crown scorch and bark thickness characteristics. They surmised, "… unmeasured root injury may have contributed to the higher than expected mortality. However, because several of the dead Douglas-firs received minimal heating, insects appear to be responsible for part of the additional mortality." Ryan and Amman (1996) showed after Yellowstone Park fires of 1988, 77% of the Douglas-fir; 61% of the lodgepole pine; 94% of the Engelmann spruce and 100% of the subalpine fir had been killed by a combination of fire injury and/or bark beetles.

Weatherby (1999) established a study to follow the fate of selected trees in two areas burned in 1994 on the Payette National Forest, Idaho. Her work illustrated the feasibility of predicting survivability based on breast-height diameter, percent crown scorch, and percent of circumference of bole (or roots) charred. In one area (French Creek), of 121 grand fir and 82 Douglas-fir monitored following the 1994 wildfire, 41% of the grand fir and 13% of the Douglas-fir had died. Of these, about half the mortality for each tree species was attributed to bark beetles. In another (Pony Creek), 36% of the Douglas-fir and 16% of the ponderosa pine had died by 1998. Bark beetles killed slightly more than two-thirds of the dead Douglas-fir (67%) and one-fourth (27%) of the dead ponderosa pines.

Burn Intensity Categories and Bark Beetle Responses
Previous post-fire evaluations in the Northern Region have varied somewhat from area to area, but most are similar to ones developed following the Little Wolf Fire (Tally Lake

Ranger District, Flathead National Forest) in 1994. Fire-affected forested areas were assigned "burn intensity" categories using aerial photographs taken soon after the fire and knowledge of pre-fire stand conditions. They were refined by post-fire surveys and field verification within burned areas. Ground-char classes were based on ones described by Ryan and Noste (1985). Burn intensity (BI) classes were as follows:

BI 1: All vegetation blackened—foliage destroyed, boles deeply charred and understory vegetation burned. Approximate distribution of ground char: Unburned 0%, Light 15%, Moderate 70%, Deep 15%.

BI 2: Stems predominantly blackened, some foliage only scorched. Understory vegetation mostly burned. Ground char: Unburned 0%, Light 25%, Moderate 60%, Deep 15%.

BI 3: Most vegetation scorched with few blackened stems; small amounts of green vegetation. Ground char: Unburned 0%, Light 40%, Moderate 50%, Deep 10%.

BI 4: Predominantly, but temporarily green with scorched or blackened areas. Ground char: Unburned 15%, Light 65%, Moderate 15%, Deep <5% (Anonymous 1996).

In order to help define the likelihood of bark beetle population buildups in those areas, Gibson (1994) made the following assessments according to identified burn intensity categories:

BI 1: Few severely burned trees will be infested by bark beetles which will later damage uninjured trees. Some may attract wood wasps (horntails, family Siricidae) or wood borers (families Cerambycidae [longhorned beetles or roundheaded wood borers] and Buprestidae [flatheaded or metallic wood borers]) but they are of little threat to adjacent green trees. Where charring has destroyed or dried the phloem, no bark beetle food remains. Even most wood borers which ultimately feed within the sapwood, require relatively fresh inner bark for newly hatched larvae. Thin-barked tree species burned to the extent that inner bark is destroyed will provide little food for insects. Thicker barked species may attract some wood-inhabiting insect species or bark beetles, depending on depth and height of charring.

BI 2: Some thicker barked species—such as Douglas-fir, western larch and ponderosa pine—may survive immediate effects of fire. In the case of Douglas-fir, however, bole scorch on more than about half of the tree's circumference will likely produce a strong attraction for Douglas-fir beetles. Large-diameter, and older ponderosa pines in this category may be attacked by western pine beetles (*D. brevicomis* LeConte), or red turpentine beetles (*D. valens* LeConte); however, outbreak development of these beetles in this situation would not be expected. Severely weakened western larch may be infested by several species of wood borers. Thin-barked species in this group—lodgepole pine, Engelmann spruce, and subalpine fir—may have been burned too severely to attract bark beetles or wood borers.

BI 3: This group likely will attract the most bark beetles. Douglas-fir in this category may be less affected, depending upon degree of bark and root collar scorch, as noted earlier. Most second-growth ponderosa pine, lodgepole pine, Engelmann spruce and subalpine fir will almost certainly be attacked by bark beetles or wood borers. Smaller diameter ponderosa pines and lodgepole pines will be infested by one or more species of engraver beetles (*Ips* spp.), other secondary bark beetles (*Pityogenes* spp. and *Pityophthorus* spp.) and wood-boring beetles. We have learned that mountain pine beetles (*D. ponderosae* Hopkins) are seldom attracted to fire-weakened trees. Engelmann spruce will be attacked by spruce beetles and subalpine fir will support populations of several beetles, the most dominant being western balsam bark beetle (*Dryocoetes confusus* Swaine).

BI 4: In this latter group, bark beetle attraction will be dependent mostly upon amount of root collar damage. Most Douglas-fir, western larch and ponderosa pines will survive and not attract beetles unless smoldering ground fires significantly damaged roots or root collars. Other tree species are more likely to be infested, even though severe damage may not be readily apparent. Observations in other burned areas have shown thin-barked trees can withstand only a small amount of damage at ground level without becoming so weakened they eventually succumb to bark beetle attacks. In these areas, it is common to find trees with little apparent bole or crown damage that have been completely girdled at the root collar.

Tree Responses to Fire and Management Alternatives
Beyond the likelihood of individual trees dying directly from fire damage, there is great interest in determining which trees are at risk of subsequently being killed by bark beetles—both dependent upon, and independent from, fire effects. Ryan and Reinhardt (1988) have described the survivability of seven coniferous species, relative to crown scorch and bark thickness. Except for ponderosa pine and grand fir, they have provided a basis for defining the probability that any particular tree would survive fire injury. As noted, however; some trees "predicted" to survive might be subsequently attacked by bark beetles. On the other hand, trees directly killed by fire, may be too severely damaged to be infested by bark beetles.

Scott et al. (2002) developed a method for determining post-fire probability of survival of several coniferous species in the Blue Mountains of Washington and Oregon that has been useful as a tree-marking guide for post-fire salvage operations. Sieg et al. (2006) reported on a multi-year study, following a series of wildfires in the West. They determined the best predictors of post-fire ponderosa pine mortality—specifically, crown scorch and consumption volume. Hood et al. (2007) demonstrated the relationship between fire-damaged Douglas-fir and subsequent attack by Douglas-fir beetles. Their model can help determine not only what fire-affected Douglas-fir may ultimately die; but more importantly, which ones are most likely to attract Douglas-fir beetles within the next year or so.

Gibson et al. (1999) documented buildups of both spruce beetle and Douglas-fir beetle populations following a wildfire, and expedient management responses used to forestall significant outbreaks on the Flathead National Forest, Montana. In most cases, timing of treatments is important. Damaged trees may be infested from shortly after fires are out (within a few days) until trees either recover or phloem becomes unsuitable (as long as

1-2 years post-burn). Some treatments, such as the use of anti-aggregative pheromones, may provide critical protection for injured trees until beetle populations decline or tree vigor improves. The availability and use of these techniques are discussed in this volume by Gillette and Munson (2009). In determining what actions may be most appropriate, an estimate of tree survivability and susceptibility to bark beetles will be essential.

Fire Survivability and Likelihood of Beetle Infestation of Common Coniferous Species in the Intermountain West

Douglas-fir: Reporting results from a multi-year, post-fire study in the Greater Yellowstone Area, Ryan and Amman (1996) showed that four years following the fires, 79% of 125 Douglas-fir in their survey plots had been attacked by one or more species of insects, and 77% were dead. Seventy-one percent of the insect attacks were by Douglas-fir beetles. Dead trees had suffered greater crown scorch and bole injury; however, trees attacked by Douglas-fir beetles had more than 50% basal girdling, ample green phloem, and less than 75% crown scorch. Beetles initially attacked severely injured trees, then attacked more lightly injured trees in subsequent years. Mortality immediately following fires occurred in trees with both severe crown scorch and bole injury. The majority of subsequent mortality, however, was found in trees with little crown injury but more than 50% basal girdling. Of dead Douglas-fir, 83% had been infested by insects. In a similar survey of fire-damaged trees in central Idaho, Weatherby et al. (1994) showed that Douglas-fir which died from fire effects had 74% crown scorch, whereas those that were killed by beetles had 39% crown scorch.

Ponderosa Pine: Burns and Honkala (1990) noted, "Survival and growth of ponderosa pine usually are little affected if 50 percent or less of the crown is scorched in a fire. Six years after a fire in Arizona, however, no poles and only 5 percent of the sawtimber-size trees were living if more than 60 percent of the crown had been destroyed. Low tree vigor and cambium damage increase the likelihood of mortality." Wagener (1961) noted that extent of fire damage in ponderosa pines was at least partly a function of time of burn. Early season fires were more damaging than ones which occurred in late summer or early autumn. Likewise, time of year greatly affected subsequent bark beetle activity; and both directly affected a tree's probability of survival. He showed young, fast-growing trees on good sites were more likely to survive than old, overmature trees on poor sites. He also noted that trees with complete crown scorch will likely survive if buds and twigs are not damaged extensively and are thus capable of producing foliage the following year. An additional criterion was damage to bark and cambium—trees with both heavy foliage scorching and moderate to severe cambium kill were more likely to die later from bark beetle attacks. Though mature ponderosa pine has thick, fairly fire-resistant bark; permanent damage and death will be influenced by amount and distribution of fuels on the forest floor and other site and stand conditions. In uneven-aged stands, injury to the cambium will vary considerably from site to site. Resultant cambium damage will greatly determine tree's survivability, and cambium killing which extends for more than a few feet up the trunk will significantly reduce a tree's probability of survival. In their study, Weatherby et al. (1994) showed that few ponderosa pines greater than 4 inches diameter-at-breast-height (d.b.h.) died if crown scorch was less than 80%. Seig et al. (2006) noted the probability of a tree's survival was predominantly associated with percent of crown scorch and amount of crown consumed; but when bark beetles and d.b.h. were considered, predictive ability increased significantly.

Lodgepole Pine: According to Burns and Honkala (1990), lodgepole pine is more susceptible to fire than Douglas-fir and some of its other associates, because of its relatively thin bark. But it is less susceptible to fire than either Engelmann spruce or subalpine fir. On the other hand, success of lodgepole pine is directly affected by the role fire plays in its regeneration. Overmature lodgepole pine's susceptibility to mountain pine beetle, a beetle-killed stand's proclivity to burn, and fire's role in opening serotinous cones, has made the lodgepole pine/mountain pine beetle/fire/stand replacement cycle a well-established relationship throughout the tree's range. Although attracted to over-mature and slow-growing individuals, mountain pine beetles infrequently colonize fire-damaged lodgepole pine. Ryan and Amman (1996) showed of 151 lodgepole pine surveyed, 62% were attacked by insects and 61% (of the total) had died. Most dead trees had been extensively girdled by fire (greater than 75% of bole circumference) and had been infested by beetles. Majority of the beetles were engraver beetles (*Ips* spp.); but a few had been infested by secondary bark beetles and wood borers. Engraver beetles preferentially attacked trees with more than 75% basal girdling, but less than 50% crown scorch.

Engelmann Spruce: Probably because of their typically wetter habitats, fewer fire-effects studies have been done in Engelmann spruce stands than many other species. In their study following the 1988 fires in Yellowstone National Park, Ryan and Amman (1994) found only 17 spruce on their plots. By 1991, however, 83% of them were dead. They noted that as might be expected for thin-barked species, mortality did not vary by tree diameter. Trees which received most apparent damage, in the form of crown and bole injury, were ones most likely to die. Sixteen of 17 trees had been more than 90% girdled by fire and 82% of them had been infested by spruce beetles. In addition, because spruce is a shallow-rooted species, slow-burning fires causing significant root damage create trees which are easily windthrown. In turn, windthrown spruce on which there is little bole charring are quite likely to be infested by spruce beetles.

Subalpine Fir: Ryan and Amman (1994) noted that subalpine fir is known for its lack of fire resistance, primarily because of thin bark. They commented, "Virtually any fire vigorous enough to scorch the bark will cause cambium injury, followed by sloughing of the dead bark." In their study they found 17 subalpine fir, all of which died following the fires. Eighty-eight percent were eventually infested by woodborers, although bark damage was initially significant enough to preclude bark beetle infestations. We have noticed, however, subalpine fir with root damage is easily windthrown, as previously noted for spruce. Such trees, with little additional bole damage, are quite susceptible to western balsam bark beetles. Beetle populations building in downed trees are then likely to infest nearby green trees not affected by fires (K.E.G. and J.N. personal observations and unpublished data).

Western Larch: Ryan and Reinhardt (1988) described conditions most often affecting tree survivability following prescribed burns. They concluded that coniferous species in the northwestern United States vary widely in their resistance to fire injury, and that deeper-rooted trees tend to have thicker bark which renders them relatively resistant to fire-related damage. Burns and Honkala (1990) recorded, "Larch develops a deep and extensive root system…" and further, "Mature larches are the most fire-resistant trees in the Northern Rockies because of their thick bark, their high and open branching habit, and the low flammability of their foliage." Mature western larch is relatively fire resistant, wind firm, and have few insect pests—particularly bark beetles—which take advantage of weakened individuals or stands. Younger larch, with thinner bark and growth habits, may be more susceptible to fire injury; especially cambial damage and crown scorch, as described by Ryan and Reinhardt (1988).

Grand Fir: Little research has been conducted on the effects of wildfire in grand fir stands; however, its morphological characteristics are similar to white fir which is rated moderate in fire resistance, becoming more resistant as it ages. In both species, fire injuries may provide entry courts for significant decay organisms (Parker et al. 2006). Burns and Honkala (1990) rate grand fir as "medium" in fire resistance—less resistant than larch, ponderosa pine and Douglas-fir; but more resistant than subalpine fir and spruce. They note that its resistance to fire is based largely on habitat. On moister sites it is readily killed by ground fires. On drier sites grand fir is more fire resistant due to deeper root systems and thicker bark which develop in those environments.

Bark Beetles and Fire Interactions in Western Conifer Forests

Little is known about the topic of bark beetle outbreaks and the likelihood or fire behavior of a subsequent fire in western forest ecosystems. Most information available on this topic comes from anecdotal information and few scientific studies. This is an issue of great relevance at the present time when we consider the extensive eruptive populations of bark beetles that we have observed in recent years. Wildland-urban interface and the proliferation of private property in these areas further exacerbate the problem as fire control operations are of utmost necessity to protect residents from personal injury and loss of assets.

As indicated above, few studies have addressed this problem. Kulakoski and Veblen (in press) indicated that a 2000 spruce beetle outbreak did not appear to influence fire extent or severity in a subsequent fire in 2000 in spruce-fir forests in northern Colorado. Bebi et al. (2003) reported that a 1940s spruce beetle outbreak in central Colorado outbreak did not affect subsequent fire susceptibility. However, Bigler et al. (2005) working is the same areas as Kulakowski and Veblen (in press) concluded that the spruce beetle outbreak slightly increased the probability of high severity fire in 2002. In Alaska, Berg and Anderson (2006) concluded that there was no relationship between spruce beetle-caused tree mortality and subsequent wildfire occurrence. Lynch et al. (2006) working in lodgepole pine after the 1988 Yellowstone Fire indicated that a 1972–1975 mountain pine outbreak increased probability of burning but a 1980–1983 mountain pine beetle had no effect. Page and Jenkins (2007) suggested rates of fire spread and intensity were higher in lodgepole pine stands currently infested by mountain pine beetles, but lower in post-epidemic stands when compared to non-infested stands. Jenkins et al. (2008) described varying fire behavior with length of time following bark beetle outbreaks. It can be seen that most of the available studies come from spruce-fir and lodgepole pine forests. Forest types such as ponderosa pine and piñon-juniper woodlands remain unaddressed. It should be mentioned that in some forest types such as lodgepole pine and spruce-fir forests, infrequent high-intensity fires are part of the ecology of these forests with bark beetles not being needed for these fires to occur. Bark beetle outbreaks, however, can and do influence both fire hazard and behavior in areas where they have occurred.

Here we present some characteristics that may influence fire[3] and how bark beetles may influence those factors using examples from a mountain pine beetle outbreak in lodgepole pine forests of north-central Colorado and from a roundheaded pine beetle outbreak, *D. adjunctus* (Blandford), in south-central New Mexico. The Colorado outbreak has been causing extensive mortality in these forests since about 2001. Mortality levels are so extensive that stands normally considered less susceptible to mountain pine beetle, less than 80 ft^2/acre, are being decimated.

Here we briefly discussed some of the changes in foliar moisture, effects in stand structure, and the accumulation of downed woody debris during and after a bark beetle outbreak. These are factors known to influence fire in forest ecosystems. We discuss some preliminary modeling efforts underway.

Foliar Moisture: Dry needles play a role in crown fires (Van Wagner 1977, Chrosciewicz 1986, Agee et al. 2002). One of the short term effects of bark beetles is altering foliar moisture caused by the simple death of the tree.

We have conducted foliage sampling of beetle-killed trees and live trees to determine foliar moisture content. Reduction in foliar moisture is evident already in the early spring and by the middle of the summer is very pronounced (Table 1). The dry needles that are on trees, in effect, lower the crown base height of the tree facilitating transition to a crown fire under a lower flame length and fire line intensity (Keyes 2006).

---

[3] Hereafter in this paper when discussing "fire" we mean likelihood of fire occurrence or potential fire behavior.

Table 1—Percent foliar moisture content in live and beetle-killed trees in 2005 at different sampling dates, Fraser Experimental Forest, Fraser, CO

Sampling Date	Live Trees	Beetle-killed trees
mid-May 2006	104	64
end-July 2006	127	9
early-December 2006	114	14

Stand Structure: Among other studies, Lentile et al. (2006), and Jain and Graham (2004) discuss how forest structure influence fire severity. Bark beetles effect changes in forest structure in a variety of ways including changing stocking levels and diameter classes of remaining live trees in the affected forest. These changes directly influence canopy bulk density and can stimulate the development of fuel ladders. In north central Colorado, about 6 years into a bark beetle outbreak, we are seeing reductions in mean tree diameters of lodgepole pine from about 20 cm down to about 12 cm (fig. 1) and in stocking levels from about 28 m^2/ha down to 9 m^2/ha (fig. 2).

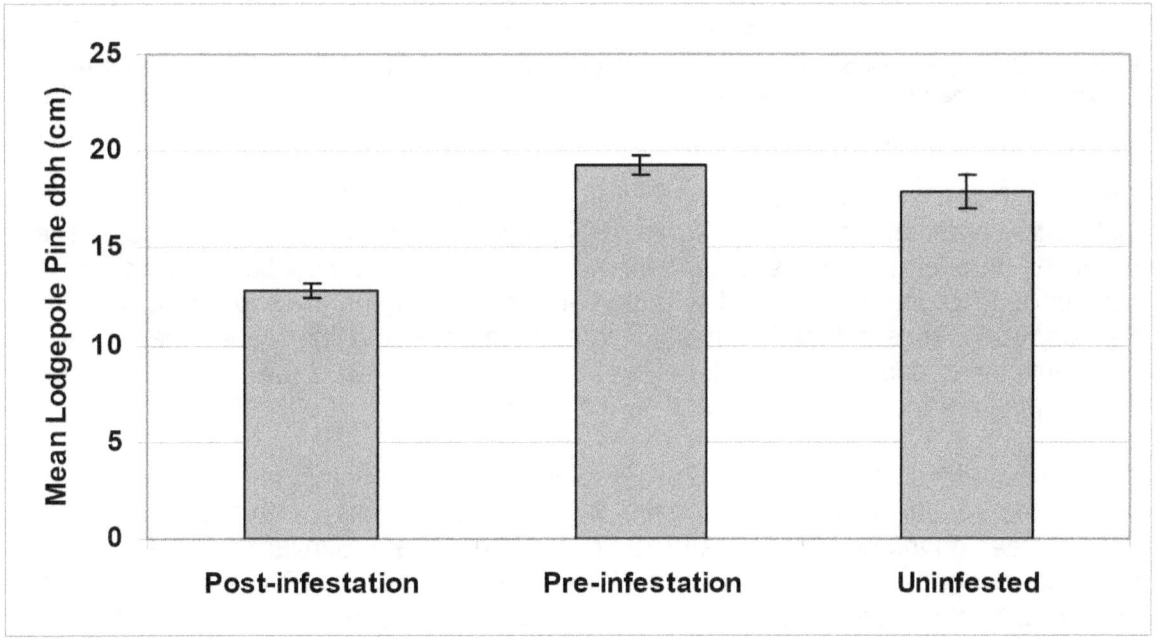

Figure 1—Mean dbh of lodgepole pine in post-infestation, pre-infestation, and uninfested, Arapahoe NF, Colorado. Error bars are standard errors.

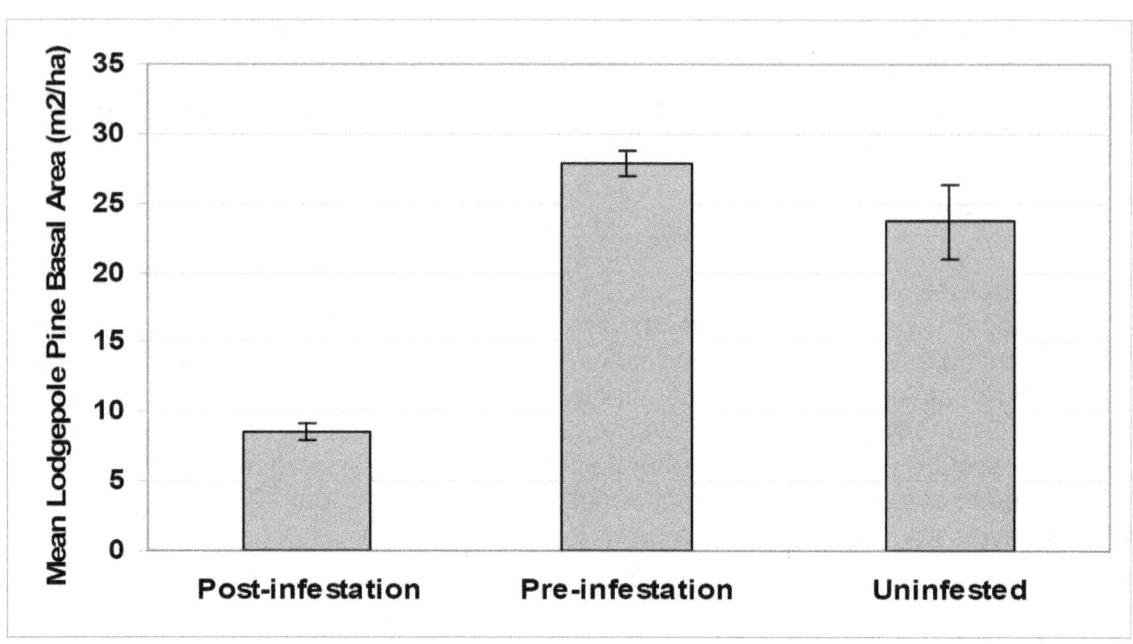

Figure 2—Basal area of lodgepole pine in post-infestation, pre-infestation, and uninfested, Arapahoe NF, Colorado. Error bars are standard errors.

Downed Wood: Another factor directly influencing fire is the type, amount, and distribution of forest fuels (Van Wagner 1977; Agee 1993). Bark beetles, through tree mortality, transform live fuels to dead fuels which will also vary spatially across the landscape following the spatial distribution of tree mortality. In the short term, less than 6 years, bark beetle-induced mortality increases the duff and litter depth, the accumulation of dead woody material less than ¼ inch but not downed wood greater than ¼ inch nor the total amount of downed woody debris.

A study by Mitchell and Preisler (1998) indicated that in an unthinned lodgepole pine stand, little tree fall occurs within the first 3 years after mortality, with about 10% and 80% of trees on the ground by 6 and 12 years, respectively. Similar fall rates have been reported for ponderosa pine, *Pinus ponderosa* (Keen 1955). The fall rate, however, can be strongly influenced by tree diameters, moisture availability in the site, and the occurrence of strong wind among others. Nevertheless the data presented by Mitchell and Preisler (1998) can be used to make some projections for the accumulation of large woody material over time. Projections made from tree mortality data by mountain pine beetle in lodgepole pine forests in Colorado result in large increases in total fuel loading 12 years after the outbreak (fig. 3). These increases in fuel accumulations may result in more intense fires with excessive soil heating.

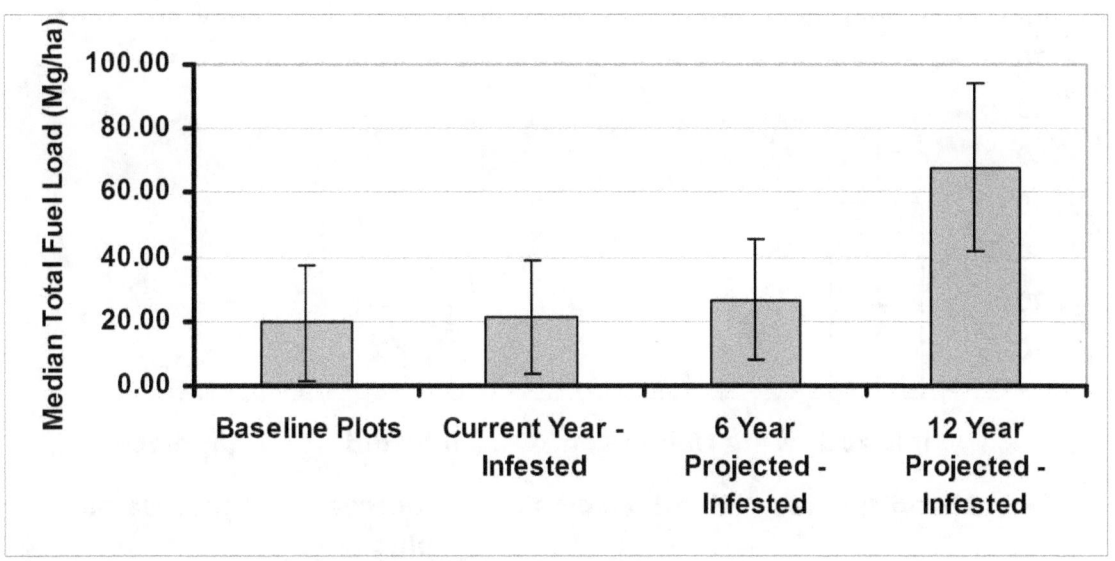

Figure 3—Median total downed woody debris accumulations in uninfested and currently infested plots, and 6- and 12-year projections, Arapaho NF, Colorado. Error bars are median absolute deviations.

During the late 1980s to early 1990s, eruptive populations of the roundheaded pine beetle caused extensive mortality in ponderosa pine. Stand susceptibility plots were established in 1994–1995. These plots were revisited 10 years after the original establishment, which represents approximately 14 years after the outbreak. Downed woody debris accumulations in mixed conifer and ponderosa pine forests increased from 6 to 40 and from 4 to 20 metric tons/hectare in mixed conifer forests and ponderosa pine forests, respectively (fig. 4).

Fire Modeling: Through preliminary modeling using the Forest Vegetation Simulator / Fire and Fuels Extention and fire models such as Behave, we have obtained projected increases in total flame length and the area affected by passive crown fires. Also obtained were decreases in crowning index, and the area affected by active crown fires. The increase in flame length may be associated with the increase in downed woody debris and the increase in passive crown fire is due to the nature of the patchy forests left after a bark beetle outbreak. A lower crowning index value means it takes a lower wind speed for fire to move within the crown. and the decrease of area affected by active crown fire may be due to the loss of crown continuity.

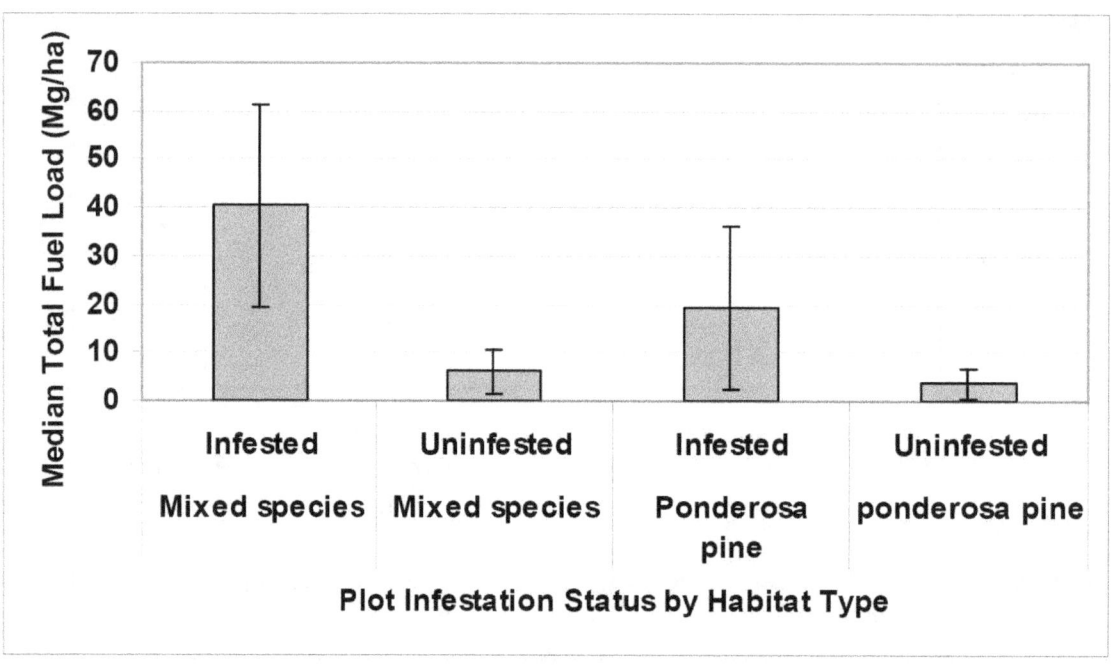

Figure 4—Median total downed woody debris accumulations in infested and uninfested plots about 14 years after a roundheaded pine beetle outbreak, Lincoln National Forest, New Mexico. Error bars are median absolute deviations.

Additional studies are underway to continue examination of downed woody debris after the occurrence of bark beetles in different forest types, to continue the study of foliar moisture dynamics, and to better clarify the changes in forest structure caused by bark beetles that can influence fire characteristics. Finally a particularly valuable resource is the availability of historical aerial detection flights of insect and disease conducted annually by Forest Health specialists and data on the location of historical fires. We are currently using GIS approaches combined with weather data to examine if and under what conditions fire may occur subsequently to a bark beetle outbreak. This information will provide the opportunity to include time since bark beetle outbreak and the occurrence of fire-conducive weather as potentially important considerations in assessing fire hazard.

Summary

We note that much remains to be learned before we will be able to accurately predict which trees will succumb to effects of a wildfire or prescribed fire, which will survive, and which of those may ultimately be killed by bark beetles. Some of the more severely affected trees will unquestionably die; some of the least affected will no doubt survive. Trees between the two extremes are ones most difficult to predict because of their varying susceptibility to bark beetles, the effects of post-fire weather, and other site/stand factors difficult to measure and not well-understood.

As previously noted, a fire-damaged tree's susceptibility to bark beetles is determined by: (1) Amount of damage and tree's response, (2) time of year fire occurs, (3) populations of bark beetles in tree's vicinity, and (4) weather for several months both pre- and post-fire.

A complex of factors is involved in any one tree's survivability. Not the least of those are pre-fire physiological condition, an array of abiotic site factors, a host of potentially damaging biotic agents, and interactions between them all. We may never unfailingly predict either post-fire survival or death for fire-damaged trees. But reasonable estimates, sufficient for most management decisions, are possible if measurable parameters are adequately considered.

Because of the area burned throughout the West since 2000, total area in the millions of acres; dealing with fire effects on all affected resources will undoubtedly extend well into the future. Yet the need to assess as quickly as possible where site rehabilitation and stabilization is most critical, and in some cases where economic values can be captured in a timely manner, will be paramount.

Bark beetles could influence subsequent fire behavior if fire occurs and depending on time since bark beetle outbreak fire and weather conditions among other factors. Reductions in foliar moisture could facilitate movement of a fire into the crown. Changes in stand structure such as reduced stocking and tree diameters can change the availability of fuel ladders, changes in understory vegetation and arrangement of fuels. Bark beetles also transform fuels from live fuels in the canopy to dead fuels on the ground. These fuel accumulations can be of significance a decade after the mortality event which will result in different fire characteristics compared to unaffected stands.

Literature Cited

Agee, J.K. 1993. Fire ecology of Pacific Northwest forests. Washington, DC: Island Press. 493 p.

Agee, J.K.; Wright, C.S.; Williamson, N.; Huff, M.H. 2002. Foliar moisture content of Pacific Northwest vegetation and its relation to wildland fire behavior. Forest Ecology and Management.167: 57–66.

Amman, G.D.; Ryan, K.C. 1991. Insect infestation of fire-injured trees in Greater Yellowstone Area. Research Note INT-398. Ogden, UT: U.S. Department of Agriculture, Forest Service, Intermountain Forest and Range Experiment Station. 9 p.

Bebi, P.; Kulakowsi, D.; Veblen, T.T. 2003. Interactions between fire and spruce beetles in a subalpine rocky mountain forest landscape. Ecology. 84: 362–371.

Berg, E.E.; Anderson R.S. 2006. Fire history of white and Lutz spruce forests on the Kenai Peninsula, Alaska, over the last two millennia as determined from soil charcoal. Forest Ecology and Management. 227: 275–283.

Bigler, C.; Kulakowski, D.; Veblen T.T. 2005. Multiple disturbance interactions and drought influence fire severity in Rocky Mountain subalpine forests. Ecology. 86: 3018-3029.

Burns, R.M.; Honkala, B.H., tech. coords. 1990. Silvics of North America, Volume 1, Conifers. Agriculture Handbook 654. Washington, DC: U.S. Department of Agriculture, Forest Service. 673 p.

Chrosciewicz, Z. 1986. Foliar moisture variations in four coniferous tree species of central Alberta. Canadian Journal of Forest Research. 16: 157–162.

Gibson, K.E. 1994. Trip Report, Unpublished office report, October 17, 1994. Missoula, MT: US Department of Agriculture, Forest Service, Northern Region, Forest Health Protection, Missoula Field Office, Federal Building, 200 E. Broadway. 3 p.

Gibson, K.; Lieser, E; Ping, B. 1999. Bark beetle outbreaks following the Little Wolf Fire, Tally Lake Ranger District, Flathead National Forest. FHP Report 99-7. Missoula, MT: U.S. Department of Agriculture, Forest Service, Northern Region. 15 p.

Gillette, N.E.; Munson A.S. 2009. Semiochemical sabotage: behavioral chemicals for protection of western conifers from bark beetle. In: Hayes, J.L.; Lundquist, J.E., comps. Western Bark Beetle Research Group—a unique collaboration with Forest Health Protection symposium, Society of American Foresters Conference, 23–28 October 2007, Portland, OR. Gen. Tech. Rep. PNW-GTR-784, Portland, OR: U.S. Department of Agriculture, Forest Service, Pacific Northwest Research Station: 85–109.

Harrington, M.G. 1996. Fall rates of prescribed fire-killed ponderosa pine. Res. Paper INT-RP-489. Missoula, MT: U.S. Department of Agriculture, Forest Service. Intermountain Research Station. 7 p.

Hood, S.; Bentz, B.; Gibson, K.; Ryan, K.; Denitto G. 2007. Assessing post-fire Douglas-fir mortality and Douglas-fir beetle attacks in the northern Rockies. Gen.Tech. Rep. RMRS-GTR-199. Fort Collins, CO: U.S. Department of Agriculture, Forest Service, Rocky Mountain Research Station. 32 p. (plus supplement).

Jain, T.B.; Graham, R.T. 2004. Is forest structure related to fire severity? Yes, no, maybe: methods and insights in quantifying the answer. In: Shepperd, W.D.; and Eskew, L.G., comps. Proceedings, Silviculture in Special Places. RMRS-P-34. Fort Collins, CO: U.S. Department of Agriculture, Forest Service, Rocky Mountain Research Station: 217–234.

Jenkins, M.J.; Hebertson, E.; Page, W.; Jorgensen, C.A. 2008. Bark beetles, fuels, fires and implications for forest management in the intermountain West. Forest Ecology Management. 254: 16–34.

Keen, F.P. 1955. The rate of natural falling of beetle-killed ponderosa pine snags. Journal of Forestry. 55: 720–723.

Keyes, C.R. 2006. Role of foliar moisture content in the silvicultural management of forest fuels. Western Journal of Applied Forestry. 21: 228–231.

Lentile, L.B.; Smith, F.W.; Shepperd, W.D. 2006. Influence of topography and forest structure patterns of mixed severity fire in ponderosa pine forests of the South Dakota Black Hills, USA. International Journal of Wildland Fire. 15: 557–566.

Lynch, H.J.; Renkin, R.A.; Crabtree, R.L.; Moorcroft, P.R. 2006. The influence of previous mountain pine beetle (*Dendroctonus ponderosae*) activity on the 1988 Yellowstone fires. Ecosystems. 9: 1318–1327.

Mitchell, R.G.; Preisler, H.K. 1998. Fall rate of lodgepole pine killed by the mountain pine beetle in central Oregon. Western Journal of Applied Forestry. 13: 23–26.

Missoula Field Office., comps. 2000. Survivability and deterioration of fire-injured trees in the northern Rocky Mountains. Forest Health Protection Report 2000–13. Missoula, MT: U.S. Department of Agriculture, Forest Service, Northern Region, State and Private Forestry. 49 p. (plus appendices).

Page, W.G.; Jenkins, M.J. 2007. Mountain pine beetle induced changes to selected lodgepole pine fuel complexes within the Intermountain Region. Forest Science. 53: 507–518.

Parker, T.J.; Clancy, K.M.; Mathiason, R.L. 2006. Interactions among fire, insects and pathogens in coniferous forests of the interior western United States and Canada. Agriculture and Forest Entomology. 8: 167–189.

Rasmussen, L.A.; Amman, G.D.; Vandygriff, J.C.; Oakes, R.D.; Munson, A.S.; Gibson. K.E. 1996. Bark beetle and wood borer infestation in the Greater Yellowstone Area during four postfire years. Res. Paper INT-RP-487. Ogden, UT: U.S. Department of Agriculture, Forest Service, Intermountain Forest and Range Experiment Station. 9 p.

Reinhardt, E.D.; Ryan, K.C. 1989. Estimating tree mortality resulting from prescribed fire. In: Baumgartner, D.M.; Breuer, D.W.; Zamora, B.A.; Neuenschwander, L.F.; and Wakimoto, R.H., comps. and eds. Symposium proceedings: Prescribed fire in the Intermountain Region. Pullman, WA: Washington State University: 41–44.

Ryan, K.C. 1982. Evaluating potential tree mortality from prescribed burning. In: Baumgartner, D.M., ed. Symposium proceedings: Site preparation and fuels management on steep terrain. Pullman, WA: Washington State University: 176–179.

Ryan, K.C. 1990. Predicting prescribed fire effects on trees in the Interior West. In: Alexander M.E.; Bisgrove G.F., tech. coords. Proceedings, 1st Interior West Fire Council annual meeting and workshop; 1988 October 24–27. The art and science of fire management. Kananaskis Village, AB. Information Report NOR-X-309. Edmonton, AB: Northern Forestry Centre: 148–162.

Ryan, K.C.; Amman, G.D. 1994. Interactions between fire-injured trees and insects in the Greater Yellowstone Area. In: Despain, D.G., ed. Plants and their environments: proceedings of the first biennial scientific conference of the Greater Yellowstone ecosystem. Technical Report NPS/NRYELL/NRTR/93XX. Denver, CO: U.S. Department of the Interior, National Park Service: 259–271.

Ryan, K.C.; Amman, G.D. 1996. Bark beetle activity and delayed tree mortality in the Greater Yellowstone Area following the 1988 fires. In: Keane, R.E.; Ryan, K.C.; Running, S.W., eds. Proceedings Ecological implications of fire in Greater Yellowstone, 1996. Birmingham, AL: International Association of Wildland Fire: 151–158.

Ryan, K.C.; Noste, N.V. 1985. Evaluating prescribed fires. In: Lotan, J.E., ed. Proceedings, Symposium and workshop on wilderness fire. Gen. Tech. Rep INT-GTR-182. Ogden, UT: U.S. Department of Agriculture, Forest Service, Intermountain Forest and Range Experiment Station: 230–238.

Ryan, K.C.; Reinhardt, E.D. 1988. Predicting postfire mortality of seven western conifers. Canadian Journal of Forest Research. 18:1291–1297.

Scott. D.W.; Schmitt, C.L.; Spiegel, L. 2002. Factors affecting survival of fire injured trees: A rating system for determining relative probability of survival of conifers in the Blue and Wallowa Mountains. Report BMPMSC-03-01. La Grande, OR: U.S. Department of Agriculture, Forest Service, Blue Mountains Pest Management Service Center, Wallowa-Whitman National Forest. 39 p.

Sieg, C.H.; McMillin, J.D.; Fowler, J.F.; Allen, K.K.; Negron, J.F.; Wadleigh, L.L.; Anhold, J.A.; Gibson, K.E. 2006. Best predictors for postfire mortality of ponderosa pine trees in the intermountain West. Forest Science. 52: 718–728.

U.S. Department of Agriculture. 1996. Environmental assessment, spruce beetle control project. U.S. Department of Agriculture, Forest Service, Flathead National Forest, Tally Lake Ranger District. 203 p.

Van Wagner, C.E. 1977. Conditions for the start and spread of crown fire. Canadian Journal of Forest Research. 7: 23–34.

Wagener, W.W. 1961. Guidelines for estimating the survival of fire-damaged trees in California. Miscellaneous Paper No. 60. Berkeley, CA: U.S. Department of Agriculture, Forest Service, Pacific Southwest Forest and Range Experiment Station. 11 p.

Weatherby, J.C.; Mocettini, P.; Gardner, B.R. 1994. Biological evaluation of tree survivor-ship within the Loman Fire boundary, 1989–1993. Report No. R4-94-06. Boise, ID: U.S. Department of Agriculture, Forest Service, Intermountain Region. 10 p.

Weatherby, J. 1999. Inter-office memo to Forest Supervisor, Payette National Forest, January 20, 1999. Boise, ID: U.S. Department of Agriculture, Forest Service, Intermountain Region. 2 p.

Some Ecological, Economic, and Social Consequences of Bark Beetle Infestations[1]

Robert A. Progar, Andris Eglitis, and John E. Lundquist[2]

Abstract

Bark beetles are powerful agents of change in dynamic forest ecosystems. Most assessments of the effects of bark beetle outbreaks have been based on negative impacts on timber production. The positive effects of bark beetle activities are much less well understood. Bark beetles perform vital functions at all levels of scale in forest ecosystems. At the landscape level they influence forest regeneration, and at the stand level they kill mature trees thus creating gaps and forest openings that are beneficial to wildlife. They also cause overall increases in forest and stand resiliency by promoting variability in sizes and ages of trees and in species compositions. The effects of bark beetles on forest ecosystems differ with beetle species, geographical location, host species, stand density and tree age. Whereas ecological consequences are normally beneficial to forest ecosystems, socioeconomic perceptions range from positive to negative. We provide several examples from western regions that illustrate ecological, economic, or social effects of bark beetle outbreaks. These examples include information on management of bark beetle outbreaks and identify research needs for the future.

Keywords: Bark beetles, beetle impact, socioeconomic perception.

---

[1] The genesis of this manuscript was a presentation by the authors at the Western Bark Beetle Research Group—A Unique Collaboration with Forest Health Protection Symposium, Society of American Foresters Conference, 23–28 October 2007, Portland, OR.

[2] **Robert A. Progar** is a Research Entomologist, USDA Forest Service, PNW Research Station, forestry and Range Sciences Laboratory, 1401 Gekeler Lane, LaGrande, OR 97850; email: rprogar@fs.fed.us. **Andris Eglitis** is an Entomologist, USDA Forest Service, R-6 Forest Health Protection, 1001 S. Emkay Dr. Bend, OR 97702; email: aeglitis@fs.fed.us. **John E. Lundquist** is an Entomologist, USDA Forest Service, R-10 Forest Health Protection and PNW Research Station, 3301 "C" St., Suite 202, Anchorage, AK 99503; email: jlundquist@fs.fed.us.

Introduction

Forest ecosystems are comprised of complex labyrinths of plant, animal, and microscopic life interacting with the abiotic environment. When these systems are in good condition from a human perspective, they perform biogeophysical functions that, from the human perspective, maximize the flow of various services and benefits for society. These services include: water and food production; regulating services like flood control and cleaning air; providing cultural, recreational and spiritual benefits; and supporting services such as nutrient cycling that maintain the conditions of life on earth (Millennium Ecosystem Assessment 2003).

Bark beetles, other insect pests, and pathogens are among the most costly of all forest disturbance agents. Combined it is estimated that they cause losses exceeding $2 billion on 20.4 million ha of forests per year (USDA 1997). Bark beetle infestations can have vast and long lasting socioeconomic and ecological consequences on our forest landscapes (Dale et al. 2001). Characterizing and quantifying these impacts on the value of forest goods and services to human society has been a puzzling problem, and remains a significant challenge to forest managers and pest specialists (Dale et al. 2001).

Ecological consequences of bark beetle infestations

Forests are dynamic and constantly changing in response to biotic and abiotic influences generally referred to as disturbances. Disturbances play significant, even critical, roles in ecosystem functioning: both natural and human-induced disturbances shape forest systems at all spatial and temporal scales by influencing their composition, structure, and functional processes. Disturbances affect succession, net primary production, nutrient and hydrological cycling, habitat partitioning, and maintenance of species diversity. From an ecological perspective, disturbance in the forest ecosystem caused by bark beetle activity is commonly viewed as "beneficial", especially when that disturbance is within its natural bounds (Samman and Logan 2000). Native insects have co-evolved with their host tree species for many thousands or millions of years and are important regulators of native systems.

At the landscape scale, some bark beetle infestations create mosaics of forest patches of various ages, densities, species compositions, and stages of succession. Larger trees with reduced vigor are especially attractive to bark beetles as sites for reproduction. At endemic levels, for instance, bark beetles beneficially remove older, larger, weaker, dominant trees, releasing understory vegetation and catalyzing stand development. Bark beetles impact structure and function (Amman 1977, Schmid and Hinds 1974), biogeochemical and hydrological cycling (Edmonds and Eglitis 1989, McGregor 1985), net primary production and maximal stand volume (Romme et al. 1986), and ecosystem species diversity and abundance (Martin et al. 2006, McMillin and Allen 2003). Bark beetles mine the wood and introduce decay fungi that accelerate decomposition, and increase nutrient release rates from fallen logs (Edmonds and Eglitis 1989); increase nitrogen mineralization and turnover; and contribute to carbon fluxes. Some studies indicate that bark beetle attacks increase stream flow (Mitchell and

Love 1973) though other studies question this. Much about the effects of bark beetles on watershed hydrology is not well-understood (e.g., McGregor 1985). Moreover, bark beetles can be more important than other disturbance agents, including fire, in modifying forest structure because of the scale of their activities (Veblen et al. 1994). The effect of disturbance scale must be considered as there are different ecological outcomes for stand-replacing versus canopy-gap-producing events (Lundquist and Negrón 2000). At the landscape scale, infestations create mosaics of forest patches of various ages, densities, species compositions, and successional stages (Kolb et al. 1999, Schowalter 2006). The beneficial ecological roles played by bark beetles have been much less studied than the negative impacts. Rather than combat bark beetles as pests, we may want to view their population swings as symptomatic of changing environmental and stand conditions and, rather than perceive the beetle as the problem, seek to address the causes of its population outbreaks.

Outbreak dynamics

Insect populations are regulated by the interactions of many factors (Schowalter 2006). At times, beetle populations erupt into outbreaks that impact large tracts of forest at the landscape level. The causes of these sudden increases in beetle population are not well known. In general, however, several factors contribute to the occurrence of an outbreak: local populations of beetles are high; a sufficient number of suitable-sized host trees are present for breeding; host vigor may be reduced, and favorable environmental conditions exist for beetle survival. Abiotic factors like climate, weather related phenomena, geographic location, or natural disturbance, also influence the development of bark beetle populations. Biotic factors like species, age, and distribution of trees, affect bark beetle population development and spread. The likelihood of an outbreak increases when many trees are stressed and their defenses are inhibited by drought or injury.

Significance of spatial scale. The effects of bark beetle outbreaks vary with spatial scale. At the individual tree scale, bark beetles cause death, deformation, and reduced or foregone growth; at the stand scale, they change species composition and forest structure; at the landscape scale, they change patterns and enhance spatial heterogeneity. Because management objectives can occur concurrently at different scales, multiple objectives can be impacted by the same disturbance event (Erdle and MacLean 1999). Relative significance of management objectives determines which of these scales is most important.

Landscape Analysis. Sometimes, whole landscapes can be altered by bark beetles, creating mosaics of forest patches of various ages, densities, species composition, and successional stages (Kolb et al. 1999, Schowalter 2006). Geospatial analyses can highlight patterns of bark beetle effects at large spatial scales, making it possible to understand ecosystem component functions and interactions that may not be apparent at smaller scales (Gamarra and He 2008). Many spatial metrics have been developed to quantify landscape patterns, but much needs to be done on correlating changes in the values of these metrics to bark beetle activity (Keane et al. 2002, Smith et al. 2002).

Socioeconomic Impacts of Bark Beetle—Direct-Use Values

Direct use values
Pearce (2001) defines direct use values as "values arising from consumptive and non-consumptive uses of the forest". The most obvious consumptive use is, of course, timber production. Less-obvious direct uses include tourism, mineral extraction, collection of pharmaceutical supplies, fuel wood harvest, and extraction of other non-timber forest products. Most bark beetle impact assessments have been based on timber production metrics.

Economic consequences
Methods and metrics used for forest pest impact assessment of direct uses are reviewed by Stark (1987). Impact is commonly characterized as number of acres affected or number of trees killed. Sometimes percent of trees infested or destroyed within individual stands is assessed. Less often, wood volume loss is calculated and, less often still, these volume estimates are converted to monetary values.

Valuation of non-timber uses
Healthy forests provide a range of values far more extensive than just those associated with timber and other exploitable resources (Chamberlain et al. 1998), and many of these resources are becoming increasingly scarce (Zhang and Li 2005). When forest health is challenged by bark beetles or other disturbances, these resources are impacted. Assessments based on timber production are inappropriate for most non-timber goods and services. Few nontimber direct uses can be adequately assessed using timber production metrics. Kline (2007) described some of the difficulties in developing metrics useful for measuring, assessing, and appraising various objectives, especially nontimber resources. Alternative value assessment techniques have been suggested for biodiversity (Nunes and van den Bergh 2001), scenic beauty (Rosenberger and Smith 1998), nontimber forest products (Chamberlain et al. 1998), and others. Buhyoff et al. (1982), Hull et al. (1984), Schroeder and Daniel (1981), and Vining et al. (1984), for example, used photos and computer generated view sheds of mountain pine beetle infested landscapes to assess impacts on scenic beauty in Colorado and Arizona. Daniel et al. (1991) conducted a similar study for spruce beetle infestation in Alaska.

Barriers to impact assessments
Several factors complicate impact estimates for bark beetles on direct use values. Some of these include:
1. Forests are usually affected by multiple disturbance agents at the same time. Pests seldom act alone usually interacting sequentially or concurrently with other disturbances, partitioning out relative impacts of co-occurring disturbances presents a significant challenge.
2. Forests grow over large heterogeneous areas, much of which is often inaccessible.
3. Forests develop over long periods of time and go through many stages of development.

4.   Forest components commonly respond to stresses and disturbances by compensatory development that mediates ecological impacts.
5.   Pest impacts may manifest themselves at different places and different times for different forest resources.
6.   Forests are managed for multiple objectives, and bark beetles have negative impacts on some resources but have no or positive impacts on others.

The factors listed above created a set of circumstances making it "extremely confusing to define forest damage" (Alfaro 1991).

Impact assessments involving multiple uses

Impact assessments based on single variables inadequately portray the changes in complex systems. Methods based on multiple variables offer promising alternatives for characterizing pest impacts on multiple objectives. Lundquist and Beatty (1999) developed an impact assessment method and used it in mixed-conifer stands in the Blue Mountains of Oregon. This method was used to show how co-occurring objectives could be both positively and negatively affected simultaneously by the same disturbance (Lundquist et al. 2002). Unfortunately, because these types of analyses are usually abstract, they are seldom easily transferred to the end user. Much more needs to be done on impact assessment and valuation and technology transfer for complex systems associated with bark beetle outbreaks.

Community based perception of risk and loss

Flint (2006) found that different communities differed in their perception of impacts and that different communities have a different "collective experience and community risk perception". Following extensive outbreaks of spruce beetle in Alaska in the mid-1990s, the collective perception of some was that the spruce beetle was a natural component of the ecosysem and that human intervention was unnecessary. Others felt that the outbreaks were a disaster that greatly impacted their communities, socioeconomically and ecologically. Still others looked on it as an opportunity to generate income by selling and/or processing the dead standing trees. The sociological aspects of bark beetle activities are largely unexplored, and Flint's (2006) result illustrate an exciting and important avenue for future research.

Existence values

Pearce (2001) lists two additional types of values: nonuse and option. Nonuse values are values associated with a willingness to pay to conserve the forest without concern for future use. The option value is based on alternative choices or options. Option values are values associated with the "willingness to pay to conserve the option of making use of the forest even though no current use is made of it..." An option is a contract that gives its holder the right, but not the obligation, to make a choice among alternatives within a specified period of time (Brigham and Ehrhardt 2002). The concept is based on real option theory (Amram and Kuatilka 1999), which is similar to financial options, differing only in that the former involves real assets rather than financial ones. The price someone is willing to pay to retain this option is its value. Determining this value is not trivial. Black and Scholes (1973) were awarded a Noble Prize for formulating an equation that calculates the value of financial options. Similarly, much work also has been done for real options.

The importance of perception and prediction
Both nonuse and option values depend on the perceived future state of the forest. Bark beetles can alter the direction and rate of stand/forest development or succession or both, and thus the perception as well as the reality of a future condition. Both nonuse and option values seem applicable to bark beetle impact assessments, and will probably become increasingly important in the not too distance future, but to date these have been little studied and their linkages to bark beetle infestations and the impacts caused by other types of disturbances is little understood.

Management Perspectives on the Consequences of Bark Beetle Infestations

The following examples of bark beetle outbreaks in various parts of the western USA show an array of impacts and identify specific needs that managers have encountered while addressing these impacts. In addition, some "success stories" are presented as examples to draw from for dealing with future outbreaks.

Mountain pine beetle— Idaho
In 2004, the Idaho Department of Lands and the USDA Forest Service formed a partnership to help private landowners in southcentral Idaho (Sawtooth National Recreation Area) deal with the effects of a severe mountain pine beetle (*Dendroctonus ponderosae*) outbreak in lodgepole pine (*Pinus contorta*). Both agencies provided technical assistance and financial cost share grants for treating forested lands. The cost share program was based on educating homeowners and contractors in the identification of beetle-infested trees and in understanding the appropriate treatment options. Treatments included removal of infested trees, thinning stands to increase resistance to bark beetles, applying a preventive carbaryl insecticide spray to high-value trees, and applying naturally occurring repellent pheromones to individual trees.

The effort to mitigate the impacts of mountain pine beetle in this one example have cost local, state, and federal land managers approximately 1.5 million dollars in implementation of a program comprising 71 projects over the 3-year period from 2004 to 2006. The program included the following actions: 28,000 beetle-infested trees removed over 486 ha; 32,000 trees sprayed with carbaryl on 400 ha; 17,000 pheromone bubble caps deployed on 243 ha; 149 ha thinned; and 18 ha replanted. The program is still ongoing and has received high praise from agency officials and the affected community. This program, and its grant coordinator, Jim Rineholt, recently received the Regional Forester's Natural Resource Stewardship Award as an excellent example of the things that can be accomplished through collaboration and a strong commitment to sharing information and technology. Two overriding research and management needs that were identified by Rineholt were:

1.	Improve the efficacy of verbenone or other bark beetle repellents in order to reduce the need for spraying carbaryl as a preventive treatment for bark beetles (This would also be useful for protecting high-elevation whitebark pine (*Pinus albicaulis*) stands where spraying is not an option).

2.	The need to make USDA Forest Service, State and Private Forestry restoration funds available to prepare and implement vegetation management plans in developed areas.

Spruce beetle—Colorado

Spruce beetle (*Dendroctonus rufipennis*) outbreaks typically occur in dense stands of mature hosts, and the resulting mortality levels are very high. High-elevation stands on the Rio Grande National Forest are currently being affected by the spruce beetle. One particular concern involves the Wolf Creek Pass Ski Area, a popular recreation site in southern Colorado. The forest around the ski area is almost entirely composed of mature Engelmann spruce (*Picea engelmannii*), and could experience major changes if the bark beetle outbreak follows its typical patterns for these kinds of stands. These changes could impact recreation and other cultural services based on the experiences at Brian Head Ski Resort in southern Utah. The forests surrounding this resort were of similar composition to those at Wolf Creek Pass. After removing beetle-killed trees at the Brian Head Resort, the ski runs were no longer well-defined. In addition, the loss of a protective tree cover increased wind through the area and led to early and rapid loss of snow owing to lack of shading. The quality of the ski experience has been compromised because of the beetle outbreak.

Pinyon ips—Arizona

During a severe drought in 2002 and 2003, millions of pinyon pines (*Pinus edulis*) on more then 800,000 ha were killed in the Southwestern US by *Ips confusus* (pers. comm. J.D. McMillan 2007). Droughts have occurred in this area in the recent past, but the outcomes were never as extreme as they were in 2003. The direct and indirect effects of this particular outbreak were far-reaching. Of special concern were the effects on Navajo and Hopi tribal lands where traditional uses of the pinyon resource may be compromised. This loss of a major conifer species also raises concerns about future production of pinyon nuts as a food source for several wildlife species. Other important issues arising from the elevated pinyon mortality include potentially greater runoff and erosion in affected areas, increased near-ground solar radiation, an increased need for preventive sprays or repellants, and irrigation of pinyons on private lands. Several stand-replacing fires (http://wildfirenews.com/archive/070704.shtml) have already occurred across portions of the beetle killed forest and there are elevated fuel loads on other portions. Several management needs arose from this bark beetle outbreak, including the following:
• stand hazard rating systems for pinyon pine under drought and changing climate scenarios
• silvicultural prescriptions for pinyon/juniper woodlands
• information on fuel loading and potential fire behavior following *Ips* outbreaks

- new or improved preventative spray alternatives, or beetle repellents for protecting high-value pinyon and juniper
- techniques to address sociological impacts on traditional Hopi and Navajo uses of the pinyon/juniper resource

It is interesting to note that not everyone viewed the extreme pinyon mortality in a negative way. Some members of the local academic community advocated no management response and viewed the loss of pinyon in former grasslands as a positive outcome.

Western pine beetle—California
A wet period from 1890 through 1960 led to the establishment of extremely dense forests throughout California. Subsequent droughts have had dramatic effects on these forests. The most recent drought in southern California (in 2003 and 2004) has resulted in the mortality of thousands of pine trees over more than 200,000 ha. Over $500 million have been spent to remove dead trees, reduce fuels, and restore these forests. There have been numerous challenges to management of the drought-related mortality caused by the western pine beetle (*Dendroctonus brevicomis*). In particular, the public lacks an understanding of the problem. They think the forests are healthy again once the particular episode of bark beetle or fire-caused mortality has run its course. Nonetheless, there have been some successes including cooperative efforts in forming fire safe councils and area safety task forces that help address local stand thinning, fuels reduction, and restoration needs.

Research and management needs in this area include a continued look at the roles of air pollution and "pest complexes" in the health of pine forests, and making that information available to USDA Forest Service, Forest Health Protection staffs and the public. Funds for thinning treatments in southern California have practically disappeared because the forests are considered recreational rather than timber-producing.

Spruce beetle—Alaska
An outbreak of spruce beetle began in the 1980s on the Kenai Peninsula and lasted for 20 years, with unprecedented levels of tree mortality. The primary concerns from forest management agencies included increased fuel loads from dead trees and rapid growth of invasive grasses (Flint and Haynes 2006). Concerns from affected communities were much broader and included various environmental and community values. This disparity of concerns was seen as a problem by the USDA Forest Service and led to a 3-year study that found communities differed in their perception of impacts (Flint 2006). In general, people acknowledged that spruce beetle is a natural component of the ecosystem, but some felt the outbreak was a disaster that greatly impacted their communities socioeconomically and ecologically. Meanwhile, others considered it an opportunity to generate income by selling or processing beetle-killed trees. The sociological aspects of bark beetle activities are largely unexplored, and Flint's (2006) results illustrate an exciting and important avenue of research. This type of study should be carried out in other settings in order to identify those areas where public and local community perceptions do not mirror those of resource managers. The resultant

increased dialogue and understanding between communities and managers will lead to better decision making and more appropriate action in response to forest disturbances.

Acknowledgments

Thanks to Don Goheen, Bruce Hostetler, Rick Kelsey and Don Scott for helpful reviews. Their time and effort have made this a much better paper.

Literature Cited

Alfaro, R.I. 1991. Pest damage and its assessment. Northwest Environmental Journal. 4: 279–300.

Alfaro, R.I. 1991. Damage assessment and integrated pest management of forest defoliators. Forest Ecology and Management. 39: 275–281.

Amman, G.D. 1977. The role of the mountain pine beetle in lodgepole pine ecosystems impact on succession. In: Mattson, W.J., ed. The role of arthropods in forest ecosystems; Proceedings, 15th International Congress of Entomology; 1976 August 19–27, Washington DC. New York: Springer-Verlag: 3–18.

Amram, M.; Kuatilka, N. 1999. Real options: managing strategic investments in an uncertain world. IEEE Engineering Management Review. 27(3): 94–102.

Black, F.; Scholes, M. 1973. The pricing of options and corporate liabilities. Journal of Political Economy. May/June: 637–659.

Buhyoff, G.J.; Wellman, J.D.; Daniel, T.C. 1982. Predicting scenic quality for mountain pine beetle and western spruce budworm damaged forest vistas. Forest Science. 28 (4): 827–838.

Brigham, E.F.; Ehrhardt, M.C. 2002. Financial management. Theory and practice, 10th ed. Mason, OH: Thomson Southwestern. 1051 p.

Chamberlain, J.; Bush, R.; Hammett, A.L. 1998. Non-timber forest products—the other forest products. Forest Products Journal. 48(10): 10–19.

Cleaves, D.A.; Brodie, J.D. 1990. Economic analysis of prescribed burning. In: Walstad, J.D.; Radosevich, S.R.; Sandberg, D.V., eds. Natural and prescribed fire in the Pacific Northwest forests. Corvallis, OR: Oregon University Press: 271–282.

Collins, S. 2005. Putting natural capital to work. Speech at Ecosystem Services Conference, 29 September 2005, Washington D.C. http://www.fs.fed.us/news/2005/speeches/05/capital-work.shtml. (12 January, 2009).

Cummings, R.G.; Harrison, G.W. 1995. The measurement and decomposition of nonuse values: a critical review. Environmental and Resource Economics. 5: 225–247.

Daily, G.D. 1997. Nature's services. Washington DC: Island Press. 392 p.

Daily, G.C.; Alexander, S.; Ehrlich, P.R.; Goulder, L.; Lubchenco, J.; Matson, P.A.; Mooney, H.A.; Postel, S.; Schneider, S.H.; Tilman, D.; Woodwell, G.M. 1997. Ecosystem services: benefits supplied to human societies by natural ecosystems. Issues in Ecology. 2: 1–16.

Daily, G.D. 2000. Management objectives for the protection of ecosystem services. Environmental Science and Policy. 3: 333–339.

Dale, V.H.; Joyce, L.A.; McNulty, S.; Nielson, R.P.; Ayres, M.P.; Flannigan, M.D.; Hanson, P.J.; Irland, L.C.; Luego, A.E.; Peterson, C.J.; Simberloff, D.; Swanson, F.J.; Stocks, B.J.; Wotton, M.B. 2001. Climate change and forest disturbances. Bioscience. 9: 723–734.

Daniel, T.C.; Orland, B.; Hetherington, J.; Paschke, J.L. 1991. Public perception and attitudes regarding spruce bark beetle damage to forest resources on the Chugach National Forest, Alaska. Unpublished report for U.S. Department of Agriculture, Forest Service, Forest Health Management, Region 10. 33 p.

Edmonds, R.L.; Eglitis, A. 1989. The role of the Douglas-fir beetle and wood borers in the decomposition of and nutrient release from Douglas-fir logs. Canadian Journal of Forest Research. 19: 853–859.

Erdle, T.A.; MacLean, D.A. 1999. Stand growth model calibration for use in forest pest impact assessment. The Forest Chronicle. 75(1): 141–152.

Flint, C.G. 2006. Community perspectives on spruce beetle impacts on the Kenai Peninsula, Alaska. Forest Ecology Management. 227: 207–218.

Flint, C.G.; Haynes, R. 2006. Managing forest disturbances and community responses: lessons from the Kenai Peninsula, Alaska. Journal of Forestry. 104: 269–275.

Gamarra, J.G.P.; He, F. 2008. Spatial scaling of mountain pine beetle infestations. Journal of Animal Ecology. 77(4): 796–801.

Hull, B.R. IV.; Buhyoff, G.J.; Daniel, D.C. 1984. Measurement of scenic beauty: the law of comparative judgment and scenic beauty estimation procedures. Forest Science. 30(4): 1084–1096.

Keane, R.E.; Rollins, M.G.; McNicoll, C.H.; Parsons, R.A. 2002. Integrating ecosystem sampling, gradient modeling, remote sensing, and ecosystem simulation to create spatially explicit landscape inventories. Gen. Tech. Rep. RMRS-GTR-92. Fort Collins, CO: U.S. Department of Agriculture, Forest Service, Rocky Mountain Research Station. 61 p.

Kline, J.D. 2007. Defining an economics research program to describe and evaluate ecosystem services. Gen. Tech. Rep. PNW-GTR-700. Portland, OR: U.S. Department of Agriculture, Forest Service, Pacific Northwest Research Station. 46 p.

Kolb, T.E.; Dodds, K.A.; Clancy, K.M. 1999. Effect of western spruce budworm defoliation on the physiology and growth of potted Douglas-fir seedlings. Forest Science. 45: 280–291.

Lundquist, J.E.; Beatty, J.S. 1999. A conceptual model for defining and assessing condition of forest stands. Environmental Management. 23: 519–525.

Lundquist, J.E.; Negrón, J.F. 2000. Endemic forest disturbances and stand structure of ponderosa pine (*Pinus ponderosa*) in the upper pine creek research natural area, South Dakota, USA. Natural Areas Journal. 20: 126–132.

Lundquist, J.E.; Goheen, E.M.; Goheen, D.J. 2002. Measuring positive, negative, and null impacts of forest disturbances: a case study using dwarf mistletoe on Douglas-fir. Environmental Management. 30(6): 793–800.

Martin, K.; Norris, A.; Drever, M. 2006. Effects of bark beetle outbreaks on avian biodiversity in the British Columbia interior: Implications for critical habitat management. British Columbia Journal of Ecosystem Management. 7(3): 10–24.

McGregor, M.D. 1985. Soil and water quality. In: McGregor, M.D.; Cole, D.M., eds. Integrating management strategies for the mountain pine beetle with multiple-resource management of lodgepole pine forests. Gen. Tech. Rep. INT-GTR-174. Ogden, UT: U.S. Department of Agriculture, Forest Service, Intermountain Research Station. 44 p.

McMillin, J.D.; Allen, K.K. 2003. Effects of Douglas-fir beetle (Coleoptera: Scolytidae) infestations on forest overstory and understory conditions in western Wyoming. Western North American Naturalist. 63: 498–506.

Millennium Ecosystem Assessment. 2003. Ecosystems and human well-being. Washington DC: Island Press. 245 p.

Nunes, P.A.L.D.; van den Bergh, J.C.J.M. 2001. Economic valuation of biodiversity: sense or nonsense? Ecological Economics. 39: 203–222.

Pearce, D.W. 2001. The economic value of forest ecosystems. Ecosystem Health. 7(4): 284–296.

Progar, R.A. 2005. Five-year operational trial of verbenone to deter mountain pine beetle (*Dendroctonus ponderosae*; Coleoptera: Scolytidae) attack of lodgepole pine (*Pinus contorta*). Environmental Entomology. 34: 1403–1407.

Romme, W.H.; Knight, D.H.; Yavitt, J.B. 1986. Mountain pine beetle outbreaks in the Rocky Mountains: regulators of primary productivity? American Naturalist. 127: 484–494.

Rosenberg, R.S.; Smith, E.L. 1998. Assessing forest scenic beauty impacts of insects and management. FHTET 98-08, Fort Collins, CO: U.S. Department of Agriculture, Forest Service, Forest Health Protection, Forest Health Technology Enterprise Team. 39 p.

Ross, S.A.; Westerfield, R.W.; Jordan, B.D. 2000. Fundamentals of corporate finance. Boston: The McGraw-Hill Companies. 694 p.

Samman, S; Logan, J. 2000. Assessment and response to bark beetle outbreaks in the Rocky Mountain Area. Gen. Tech. Rep. RMRS-GTR-62. Fort Collins, CO: U.S. Department of Agriculture, Forest Service, Rocky Mountain Research Station. 46 p.

Schmid, J. M.; Hinds, T. E. 1974. Development of spruce-fir stands following spruce beetle outbreaks. Res. Pap. RM-131. Fort Collins, CO: U.S. Department of Agriculture, Forest Service, Rocky Mountain Research Station. 16 p.

Schroeder, H.W.; Daniel T.C. 1981. Progress in predicting the perceived scenic beauty of forest landscapes. Forest Science. 27(1): 71–80.

Schowalter, T.D. 2006. Insect ecology: an ecosystem approach. 2nd ed. San Diego: Academic Press. 572 p.

Smith, A.H., Pinkard, E.A.; Stone, C.; Battaglia, M.; Mohammed, C.L. 2005. Precision and accuracy of pest and pathogen damage assessment in young eucalypt plantations. Environmental Monitoring and Assessment. 111: 243–256.

Smith, E.L.; McMahan, A.J.; Eager, T. 2002. Landscape analysis applications of the westwide pine beetle FVS extension. In: Crookston, N.L.; Havis, R.N., comps. Proceedings of the second Forest Vegetation Simulator conference; 2002 February 12–14, RMRS-P-25. Fort Collins, CO: U.S. Department of Agriculture, Forest Service, Rocky Mountain Research Station: 62–68.

Stark, R.W. 1987. Impacts of forest insects and diseases: significance and measurement. Critical Reviews in Plant Sciences. 5: 161–203.

US Department of Agriculture. 1997. Forest insect and disease conditions in the United States 1996. Washington DC: U.S. Department of Agriculture, Forest Service, Forest Health Protection. 87 p.

Walsh, R.G.; Loomis, J.B.; Gillman, R.A. 1984. Valuing option, existence, and bequest demand for wilderness. Land Economics 60(1): 14–29.

Veblen T.T.; Hadley, K.S.; Nel, E.M.; Kitzberger, T.; Reed, M.; Villalba, R. 1994. Disturbance regime and disturbance interactions in a Rocky Mountain subalpine forest. Journal of Ecology. 82: 125–135.

Vining, J.; Daniel, T.C.; Schroeder, H.W. 1984. Predicting scenic beauty in forested residential landscapes. Journal of Leisure Research. 16: 124–135.

Werner, R.A.; Holsten, E.H. 1984. Factors influencing generation times of spruce beetles in Alaska. Canadian Journal of Forest Research. 15: 438–443.

Wood, S.L. 1982. The bark and ambrosia beetles of North and Central America (Coleoptera: Scolytidae), a taxonomic monograph. Great Basin Naturalist Memoirs. 6. 1359 p.

Zhang, Y.; Li, Y. 2005. Valuing or pricing natural and environmental resources? Environmental Science & Policy. 8: 179–186.

WESTERN BARK BEETLE

RESEARCH GROUP

Semiochemical Sabotage: Behavioral Chemicals for Protection of Western Conifers From Bark Beetles[1]

Nancy E. Gillette and A. Steve Munson[2]

Abstract

The discovery and elucidation of volatile behavioral chemicals used by bark beetles to locate hosts and mates has revealed a rich potential for humans to sabotage beetle host-finding and reproduction. Here, we present a description of currently available semiochemical methods for use in monitoring and controlling bark beetle pests in western conifer forests. Delivery systems include hand-applied methods, such as semiochemical-releasing bubblecaps, pouches, and "puffers," as well as products that can be applied by aircraft such as semiochemical-releasing flakes. Descriptions of both attractant-based ("pull") and anti-attractant-based ("push") strategies are provided. Examples are provided for the major bark beetle pests in western North America, including the mountain pine beetle (*Dendroctonus ponderosae* Hopkins), western pine beetle (*Dendroctonus brevicomis* LeConte), the Douglas-fir beetle (*Dendroctonus pseudotsugae* Hopkins), the spruce beetle [*Dendroctonus rufipennis* (Kirby)], and the red turpentine beetle (*Dendroctonus valens* LeConte),.

Keywords: Pheromones, allomones, kairomones, IPM, trap-out, trap trees, push-pull, pine, Douglas-fir, spruce.

[1] The genesis of this manuscript was a presentation by the authors at the Western Bark Beetle Research Group—A Unique Collaboration with Forest Health Protection Symposium, Society of American Foresters Conference, 23–28 October 2007, Portland, OR.

[2] **Nancy E. Gillette** is a Research Entomologist, USDA Forest Service, Pacific SW Research Station, 800 Buchanan Street, Albany, CA 94710: e-mail ngillette@fs.fed.us. **A. Steve Munson** is an Entomologist, USDA Forest Service, Forest Health Protection, 4746 S. 1900 E., Ogden, UT 84403; e-mail smunson@fs.fed.us.

Introduction

Background

Bark beetles are the most damaging insect pests of conifer forests in western North America (Furniss and Carolin, 1977) and outbreaks are increasing (Hicke et al. 2006, Hicke and Jenkins 2008, Logan and Powell 2001). For example, a current epic outbreak of mountain pine beetle in British Columbia, Canada, has affected over 9.2 million hectares of ponderosa pine (*Pinus contorta* Dougl.) (Westfall 2007) and has breached the Continental Divide, spilling over into interior Canada (Wilent 2005). This bark beetle outbreak is the largest ever documented, and is expected to continue until either the host is depleted or severe cold weather reduces beetle populations (Ebata 2004). Outbreaks of this magnitude have the potential to convert large regions of boreal and temperate forest from carbon sinks to carbon sources, exacerbating global warming (Kurz et al. 2008a, 2008b). The MPB could infest millions of hectares of jack pine (*Pinus banksiana* Lamb.) in the vast boreal forests of Canada and the north central United States, and climate change may favor *D. ponderosae* range extensions into this habitat (Carroll et al. 2003, Logan and Powell 2001, Ono 2004). Heavily stocked or old growth stands are particularly at risk (Shore et al. 2000, Wood et al. 1985), with extensive outbreaks predicted for many locations in the western United States (Krist et al. 2007). Forest managers have therefore sought methods to mitigate the effects of these pests. To this end, efforts have focused on the development of better methods to prevent losses of forest trees to bark beetle outbreaks, particularly high-value trees in the urban-interface, recreation areas, and high elevation ecosystems.

Semiochemical-based bark beetle control has been the subject of a substantial research effort (summarized by Borden 1997, Skillen et al. 1997, and Wood et al. 1985) since the identification of the first bark beetle pheromones (Silverstein et al. 1966, 1968). Land managers have had high expectations for the development of pheromones and other behavioral chemicals for bark beetle control because of limitations encountered with other pest control methods. For example, it is widely accepted that maintenance of stand health and vigor through vegetation management is the most durable approach to "beetle-proofing" stands (Amman et al. 1991; Amman and Logan 1998; Fettig et al. 2006c, 2007; Negrón et al. 2001; Whitehead and Russo 2005), but management objectives sometimes require maintenance of high basal area (Andrews et al. 2005) and/or the creation of down woody material that increases stand susceptibility to bark beetle attack (Ross et al. 2006). Treatments to reduce stand density are also time-consuming and can incur regulatory obstacles that may delay the implementation of treatments until stands have already been compromised by bark beetle attacks. Sanitation and salvage may help mitigate the effect of bark beetles, particularly in small, isolated infestations (Bentz and Munson 2000), but these methods are often insufficient and/or of unproven efficacy for landscape-altering outbreaks. Biological control, while generally a desirable approach to pest management, is of limited use against native bark beetle pests using their native natural enemies. While biological control manipulations such as augmentation of native natural enemies or inundative release of parasitoids and predators are theoretically possible, it is unlikely that they would be implemented over large scales because of logistical constraints. Insecticides have been tested for decades for bark beetle control (DeGomez et al. 2006; Fettig et al. 2006a, 2006b, Haverty et al. 1998; Naumann and Rankin 1999), but they are generally too toxic, time-consuming, expensive, and difficult to deploy in remote areas for widespread use on public lands, with the exception of high-value trees in the wildland-urban interface, campgrounds, ski resorts, and administrative sites. The development of semiochemicals, therefore, is an appealing alternative to other integrated pest management (IPM) methods for mitigation of damage by bark beetles. IPM is a systematic approach to pest control that incorporates monitoring to assess the need for treatments, then initiates treatments as needed, beginning with the most environmentally benign methods. Typically, cultural or mechanical control methods are attempted first, followed by biological control and/or semiochemicals, then use of insecticidal control only if other methods fail (Kogan 1998, Smith 1962).

Early attempts to control damage by bark beetles using semiochemicals were handicapped by insufficient information about the components of the semiochemical blends and by inadequate release devices. That is, the release devices either did not release sufficient quantities of semiochemicals or did not release the semiochemicals long enough to protect stands during the entire flight periods of the targeted pest species (Holsten et al. 2000). Because of the limitations of other pest control strategies and the urgent need to protect conifers from bark beetle attack, recent research has focused on the development of more effective active ingredients such as aggregation pheromones, synergists, and anti-attractants and on more effective release devices for dispersal of these semiochemicals. New information about behaviorally active semiochemical blends, newer release devices, and the integration of semiochemicals with silvicultural pest management methods have led to more effective strategies to minimize damage by these pests.

In describing case histories of semiochemical methods for controlling western bark beetles, we have organized the discussion by pest species. Although we discussed southern pine beetle (*Dendroctonus frontalis*) applications in our symposium presentation (Clarke et al. 1999, Salom et al. 1995), in keeping with the overall symposium theme, this article will be restricted to the major western bark beetle species. Likewise, we have not included discussions of the use of semiochemicals for monitoring invasive bark beetle species (see Seybold and Downing, this Proceedings) or for the control of ambrosia beetles or forest Lepidoptera, although the use of sex pheromones in mating disruption has been quite successful for reducing damage by forest moths. The resources described below are not intended as an exhaustive list; this is an active field of research and development, with new active ingredients and release systems being constantly developed and tested for efficacy.

Semiochemicals and Applied Chemical Ecology
Semiochemicals are chemicals emitted by one organism that can affect the behavior of another organism; the term "semiochemical" is derived from the Greek "semeion," meaning signal. The terminology for describing semiochemicals has changed over time, with multiple terms for the same phenomena (Nordlund and Lewis 1981). Terms used in the past, with some overlap in meaning, include

- Infochemicals
- Signalling chemicals
- Behavioral chemicals
- Behavior modifying chemicals
- Pheromones
- Semiochemicals

The term "semiochemical" has been widely accepted as an umbrella term for these chemicals. Semiochemicals that act within a species are called pheromones, and those that act between species are referred to as allelochemicals (fig. 1). Allelochemicals that benefit the sending organism are called allomones (from the Greek "allos," other), and those that benefit the receiving organism are called kairomones (from the Greek "kairos," opportunist). Those that benefit both the sender and receiver are called synomones.

For example:
- Bark beetles use aggregation pheromones to concentrate enough adult beetles of the same species to overcome tree defenses (acts within a species to enhance progeny survival).
- Humans infected with malaria exhale volatile allelochemicals that attract the Anopheline mosquito vectors of malaria (acts between species to the detriment of the human host but to the benefit of both the mosquito and the malaria parasite).
- Skunks use a noxious spray to repel predators (benefits the sender, thus an allomone).
- Ambrosia beetles use ethanol emanating from fermenting tree tissues as a cue in host location (benefits the receiver, thus a kairomone).

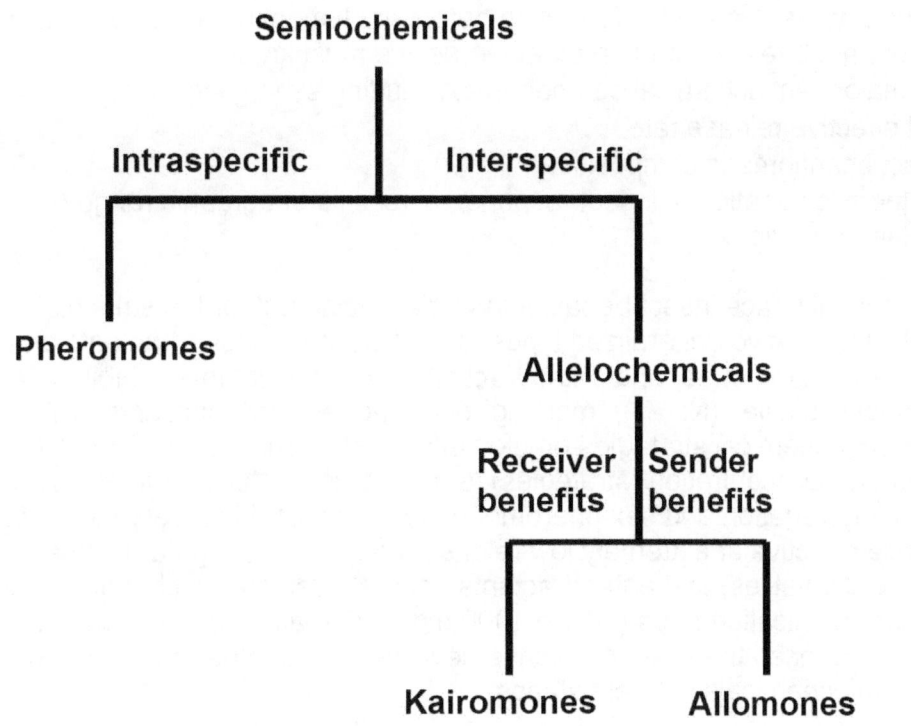

Figure 1—Diagram of semiochemical activity.

In practice, most semiochemicals used operationally in pest control are either pheromones or kairomones. There are several other issues that are important to keep in mind when using semiochemicals:

- Most semiochemicals are multifunctional
  - Their release rate can affect the behavior elicited
  - They can be attractive at low rates, repellent at high rates
- Most semiochemicals are multicomponent blends
  - The components of the blend may be inactive by themselves
  - Many aggregation pheromone blends include host volatile compounds with the beetle-produced pheromones, often as synergists
- Chiral pheromones and kairomones
  - Many semiochemicals are optically active and can exist in "mirror image" forms (enantiomers, "plus" vs. "minus," "*R*" vs. "*S*," or "L" vs. "D"), which have nearly identical physical properties but can result in different behavioral responses by the receiving insect
  - The "antipode" or opposite enantiomer of a semiochemical, for example, may be inactive or may even interrupt the response to the other enantiomer
- Insects can use different semiochemical "dialects" in different parts of their range
  - Therefore it is important to use semiochemicals that are regionally appropriate
  -

It is therefore crucial to have certain information before implementing a semiochemical-based strategy for bark beetle control. In other words, we must know

- All of the major semiochemical components, including synergists
- The most effective release rate
- The correct enantiomeric composition
- Whether there is variation in insect response across its geographic range (i.e., we need the right "dialect")

Semiochemicals can influence insect behavior in myriad ways, but for the sake of simplicity we will treat just two generalized types: attraction (e.g., host attractants and aggregation and sex pheromones) and anti-attraction (e.g., interruptants, inhibitors, anti-aggregants, non-host volatiles (NHVs), "marking" pheromones, and repellants). All of the widely used semiochemical strategies employ attractants ("pull," "attract-and-kill," and "containment-and-concentration" strategies), anti-attractants ("push" strategy), or both ("push-pull"). Aggregation and sex pheromones typically provide a very strong cue, and they are hence effective at extremely low release rates (1 to 10 mg/day). Other attractants (e.g., host volatiles) and anti-attractants generally require much higher release rates and/or application rates (100 to 1000 mg/day) to affect beetle behavior. These traits have influenced the types of release devices that have been developed for the dispersal of semiochemicals in forest stands.

Commonly Used Semiochemical-Based Strategies
- *Monitoring* is not intended to control bark beetle populations, but to detect and measure population levels of bark beetles using attractants (usually aggregation pheromones) in release devices such as bubblecaps, vials, or solid polymer tubing
- *Trap-out* removes bark beetles from the population by luring them with attractants released from bubblecaps, vials, or solid polymer tubing. These techniques include traps, trap-trees and attract-and-kill
- *Repellency* (interruption or inhibition of aggregation or host location) causes dispersal away from stands using repellents in release devices such as bubblecaps, pouches, puffers, or flakes
- *Push-pull* involves the use of an attractive pheromone at the perimeter of stands coupled with a repellent pheromone in the center of treated stands. This technique, combining both trap-out and repellency (Cook et al. 2007), has been shown to improve efficacy of repellents in some cases

Terminology and techniques
Trap "lures" normally consist of aggregation pheromones combined with attractant or synergistic host volatiles (Seybold et al. 2006), and are meant to be attached to multiple-funnel, panel, or vane traps (fig. 2). Tree "baits," on the other hand, consist of aggregation pheromones formulated without the host volatiles and are intended to be stapled or nailed to the host tree trunk. The host tree is presumed to release the monoterpene synergists. In some cases, host monoterpenes synergize the attraction of aggregation pheromones and are thus considered part of the pheromone blend.

A.  B.

Figure 2—A, multiple funnel trap (reprinted with permission from Pherotech International (now Contech International)); B, panel trap (reprinted with permission from Aptive, Inc.).

Non-host volatiles (NHVs), which include green leaf volatiles (GLVs) and angiosperm volatiles (i.e., non-conifer volatiles, collectively), have shown promise in increasing the efficacy of one of the two primary anti-attractants, verbenone, for some beetle species. The effective blend is often quite species-specific, so a single blend will probably not serve all needs.

Release devices such as bubblecaps, pouches, puffers, and vials range in size from about 2.5–10.2 cm and are meant to be manually attached to the substrate (e.g., traps or trees) (fig. 3A–C). Bubblecaps, pouches, vials, and flakes are "passive" releasers, so their release rate varies with changes in temperature and humidity. In practice these variations may not be important, because temperature changes also affect insect emergence and flight, often in ways that parallel the need for semiochemical emission. Puffers are small battery-activated reservoirs that emit frequent, measured puffs of semiochemical, thus overcoming the problem of depletion of the release device and variable release rates under fluctuating temperatures. Flakes are much smaller, usually 3–6 mm^2 in size, and are intended for aerial application over large areas. They can be applied dry, so that they fall to the forest floor, or with a liquid sticking agent that makes them adhere to the forest canopy. Flakes can also be applied using a hand-held fertilizer spreader to cover smaller acreages. Flakes, like other passive releasers, are temperature-dependent in their release profiles.

A.

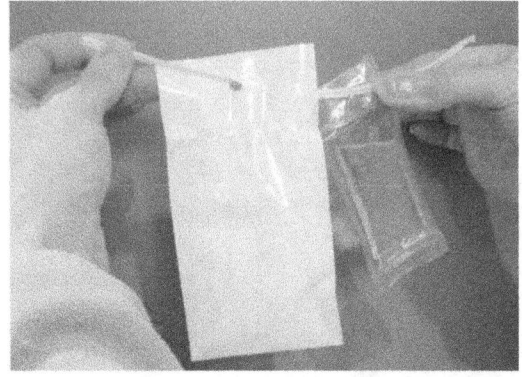

B.

C.

Figure 3A, DFB two-part lure; 3B, MCH bubblecap; 3C, verbenone pouch (all with permission of Synergy Semiochemicals).

Baited traps
Baited traps are typically used to determine flight periodicity in order to time the implementation of suppression projects. Baited traps can also be used as a suppression tactic, in which sufficient numbers of insects are trapped to reduce local infestation levels. This tactic is often combined with other suppression treatments to enhance treatment success. When used for suppression, baited traps should be placed at least 25 meters from susceptible hosts, and generally in an elevated and/or shaded position. Multiple-funnel traps (with varying numbers of funnels) or panel traps (fig. 2 A-B) are both effective for monitoring bark beetles.

Trap trees for concentration or trap-and-kill

When used as a suppression tactic (concentration or trap-and-kill), baited trees should be of fairly large diameter and in shaded sites. Adjacent hosts may also be attacked, so it is important to place baits carefully to avoid undesired tree mortality. All attacked trees are intended to be sacrificed, and once they are infested they should be removed, burned, or debarked.

Aerially applied flakes

Semiochemical-releasing flakes have been used for decades in the Gypsy Moth Slow-the-Spread program (Sharov et al. 2002), but have been only recently developed for bark beetle pheromones (Gillette et al. 2006, 2009a, 2009b). Recent tests have demonstrated the promise of this technology for control of Douglas-fir beetle and MPB, and testing continues for other bark beetle species.

Semiochemicals for Major Western Bark Beetle Pests

Mountain pine beetle (MPB)

Effective techniques have been developed for most of the major hosts of MPB, including lodgepole pine, whitebark pine *Pinus albicaulis* Engelm.), limber pine (*Pinus flexilis* James), and ponderosa pine. The primary anti-attractant for MPB, verbenone, has also shown behavioral activity for several other bark beetle species and is produced by a wide variety of organisms including bacteria, fungi, gymnosperms and angiosperms (Gillette et al. 2006). Combining verbenone with nonhost volatiles may provide better protection than verbenone alone (Huber and Borden 2001).

Monitoring and Trapping (Pull)

A blend of *trans*-verbenol, *exo*-brevicomin, myrcene, and terpinolene is highly effective for attracting MPB when used as a trap lure. Earlier research suggested that the first three components comprised the aggregation attractant blend (Borden and Lacey 1985, Conn et al. 1983), but more recent work has shown that the addition of terpinolene greatly increases trap catch (Pureswaran and Borden 2005). If reduced attraction is desirable, for example where there is a risk of inducing attack on adjacent healthy trees, the two-component tree bait (*trans*-verbenol and *exo*-brevicomin) can be deployed instead (Borden et al. 1993). Attract-and-kill or concentration techniques have been tested for decades and were shown to be effective in reducing rate of attack on adjacent trees (Gray and Borden 1989, Smith 1986). The four-component aggregation semiochemical blend described above is presumably optimal for trapping-based methods. The earliest trap-based control methods utilized insecticide-treated trees that were baited with the aggregation pheromone (Smith 1986). Vandygriff et al. (2000) successfully used aggregation pheromones to focus beetle attacks in areas designated for fuelwood harvest, potentially improving stand health in baited sites. More recent studies have shown good control of adjacent stands by baiting "sacrificial trees" that are intended for immediate harvest as soon as they are attacked and fully colonized (Borden et al. 2003, 2006, 2007).

Push

The interruptant verbenone has been widely tested for repellency of MPB. Early tests using lower-release rate bubblecapsules did not provide sufficiently high release (Holsten et al. 2000, Lister et al. 1990), but subsequent higher-release devices called pouches (Contech International, formerly Pherotech International, Delta, BC, Canada; Synergy Semiochemicals, Burnaby, BC, Canada; ChemTica USA, Durant, OK, USA; Aptiv, Portland, OR, USA; Alpha Scents, Bridgeport, NY, USA) generally have provided significant protection (Bentz et al. 2005; Borden et al. 2004, 2007; Gibson and Kegley 2004; Kegley et al. 2003; Kegley and Gibson 2004; Progar 2003). In some cases of extreme beetle pressure and/or stand susceptibility, efficacy appears less certain (Progar 2005), but newer formulations are registered to allow higher application rates, which may improve efficacy (Gillette et al. 2009a). The verbenone pouches contain 7.1–7.4 g verbenone (Pherotech International, Synergy Semiochemicals). The addition of NHVs to verbenone often improves efficacy of the repellent (Borden et al. 2003, 2006, Huber and Borden 2001), but in many cases sufficient efficacy is achieved with verbenone alone (Kegley and Gibson 2004, Kegley et al. 2003). Pouches are typically applied 3–4 m above the ground and are applied to the north sides of trees in a grid with roughly 50–100 pouches per hectare, with higher rates recommended for more challenging situations. Some verbenone treatments are applied at the rate of 50 pouches/hectare with replacement at mid-season. This approach is especially desirable where weather conditions indicate that pouches may become depleted before the end of the season. Area protection treatments using verbenone are significantly more effective if all the infested trees within the treatment area are removed before beetle flight. Increasing the verbenone grid to include a 25–30 m treated buffer may also enhance efficacy. Where individual trees, rather than stands, are intended to be protected, pouches are applied at the rate of two per tree on the northeast and northwest sides of the trees. In the case of whitebark pines, which often occur as mixed stands with other pine species, adequate protection can be achieved by placing pouches on both the whitebark pines and surrounding trees, to create an area effect that ensures that the pheromone plume encompasses the trees to be protected regardless of wind direction. Additional studies are underway to test ways of increasing the efficacy of this technique, particularly by adding NHVs to the anti-attractant verbenone.

Verbenone-releasing flakes, which can be applied to individual trees using hydroseeders or to stands using aircraft or broadcast spreaders, have recently been shown to provide good protection when applied at the rate of 15 g/tree (individual tree tests, described in Gillette et al. 2006) or 370 g/hectare (aerial application tests, described in Gillette et al. 2009a).

Push-pull

Combining anti-attractants along with aggregation pheromones deployed in trap trees has been shown to provide increased protection of lodgepole pine trees from attack by MPB, with the caveats that the density of lodgepole pines should be greater than 400 stems/hectare, the mean diameter at breast height (dbh) should be equal to or less than 25 cm, the current attack rate should be less than 15%, and the tactic should be combined with sanitation harvesting to remove infested trees (Borden et al. 2006, Lindgren and Borden 1993). One study, however, questioned the need for use of the anti-attractant (Vandygriff et al. 2000), and this hypothesis warrants further examination considering the costs of deploying the anti-attractants. Vandygriff et al. (2000) showed that baiting with the attractant was highly effective in removing sufficient numbers of beetles to reduce rate of attack in treated stands as compared to controls. They also demonstrated the utility of using the tree-baiting technique as a simultaneous sanitation effort, where mistletoe-infested stands were targeted for baiting and subsequent harvest, in order to remove both the bark beetles and mistletoe inoculum.

Douglas-fir Beetle (DFB)

The DFB often builds up high populations in wind- and avalanche-thrown Douglas-fir [*Pseudotsuga menziesii* (Mirb.) Franco] trees or in fire-damaged stands (Furniss and Carolin 1977). It can be desirable to treat such areas to prevent population build-up and infestation of healthy adjacent stands (Furniss et al. 1981, 1982). The development of semiochemical methods for control of DFB has been one of the signal success stories in the history of semiochemical research and development, perhaps because DFB is reputed to be such an olfactory specialist (Campbell and Borden 2006), i.e., it relies more on olfactory cues than do some bark beetle species, and thus be more readily manipulated with semiochemicals.

Monitoring and trapping (Pull)
Seudenol (3-methylcyclohex-2-en-1-ol) or MCOL (1-methylcyclohex-2-en-1-ol), with or without frontalin and ethanol, provides excellent efficacy for trapping DFB when used with multiple funnel traps, which are reported to work better than panel traps for this beetle species (Ross and Daterman 1998). Frequent lure replacement (every 4-6 weeks) may be necessary to maintain constant levels of release.

Push
The anti-aggregation pheromone methylcyclohexenone (3-methylcyclohex-2-en-1-one or MCH) is extremely effective with several different release devices. Bubblecap release devices deployed at the rate of about 75–100/hectare to standing trees or wind- or avalanche-thrown trees have been used for decades with good success for relatively small areas, particularly in recreation sites or administrative areas (Ross and Daterman 1994, 1998; Ross et al. 1996, 2002). Individual high-value trees can be effectively protected with the application of two bubblecaps per tree. The primary limitations to the use of bubblecaps or verbenone pouches are the cost of labor for hand application and the inability to treat remote or steep terrain by hand. For these reasons, there have been several attempts to develop aerially applied products for treatment of large, remote, and/or steep areas. In the past, aerially applied granular controlled-release formulations were successful in area-wide tests (Furniss et al. 1981, 1982), and newer flake formulations (Hercon Environmental, Emigsville, PA) are showing similar promise for treatment of large areas using fixed wing aircraft or helicopters (Gillette et al. 2009b). Initial tests provided good results with 370 g of MCH/hectare, and preliminary results from ongoing tests suggest that lower application rates may provide equivalent

protection (Constance Mehmel, USDA Forest Service, Wenatchee, WA, personal communication).

Push-pull

When beetle populations are very high, stands are extremely stressed, or windstorms, avalanches, or fire have resulted in many dead or damaged trees for beetle population build-up, it is probably advisable to combine the repellent technique with a trap-out technique (Ross et al. 1994, Blackford, 2007). In this scenario, the healthy stands are treated with MCH-releasing bubblecaps or flakes, while the perimeter, especially near fallen or damaged trees, is treated with 12-funnel traps baited with the three-component lure [Seudenol (or MCOL), frontalin, and ethanol]. Care must be taken, however, to place baited traps far enough from healthy trees to avoid spill-over attack from beetles attracted to the baited traps.

Spruce Beetle (SB)

The SB normally attacks only weakened or windthrown spruce trees. Occasionally, however, large outbreaks develop in which healthy trees of all ages and diameters are attacked and killed (Furniss and Carolin 1977). The principal hosts are *Picea engelmannii* Parry, *P. glauca* (Moench) Voss, and *P. sitchensis* (Bong.) Carr.

Monitoring and trapping (Pull)

The SB is effectively attracted by either a two-component (frontalin + α-pinene) or three-component (frontalin + α-pinene + MCOL) lure, with substantial increases obtained with the addition of MCOL (Ross et al. 2005). Werner et al. (1988) used baited trap trees that were treated with a silvicide and removed from the forest to reduce populations of SB and achieve a measure of damage control for experimental purposes. However, available silvicides are not registered in the United States for this use.

Push

MCH and green leaf volatiles have been tested with some success for interruption of host location by SB (Poland et al. 1998, Werner et al. 1988), but the use of semiochemicals in a "push" strategy has only recently been shown to be successful for tree protection, probably because of the difficulty in achieving sufficient and/or sustained release in the cooler high elevation and sub-boreal regions where spruce beetle occurs (Borden et al. 1996, Holsten et al. 2000, Ross et al. 2004). Recently a type of puffer known as the Med-E-Cell, which is an active, battery-operated, timed-release device, was shown to provide significant protection for Lutz and Sitka spruce in Alaska (Holsten et al. 2003). However, other studies in Utah using MCH in the same releaser were not effective because the devices leaked and were not capable of retaining enough MCH to ensure efficacy throughout the beetle's flight period. Further studies and product development are therefore required to achieve consistent repellency of SB with this technology.

Western Pine Beetle (WPB)

The aggregation pheromone blend for WPB has been known for nearly four decades (Bedard et al. 1969, Browne et al. 1979, Silverstein 1968, Wood 1972, Wood et al. 1970) and an early trap-out study showed significant success in reducing beetle populations in ponderosa pine (*Pinus ponderosa* Laws.) stands (Bedard and Wood 1981, DeMars et al. 1980). Efforts to develop a fully operational methodology for semiochemical control of WPB has been somewhat stalled, however, probably for lack of a sufficiently effective anti-attractant semiochemical blend to deploy as a repellent strategy. Although verbenone showed some early promise as an anti-attractant for WPB (Bedard et al. 1980, Tilden et al. 1985), when used alone for tree protection its efficacy has been equivocal (Bedard and Wood 1981, Gillette et al. 2009a, 2009b). More recently, Erbilgin et al. (2007b, 2008) and Fettig et al. (2005, 2008a, 2008b) have demonstrated efficacy of adjuvants to verbenone and other active ingredients to enhance efficacy of a "push" or "push-pull" technique for WPB. The adjuvants (NHVs), which are largely those that have shown efficacy for MPB, are still being tested for area-wide use but have shown substantial efficacy in individual tree tests (Fettig et al. 2008a, 2008b).

Monitoring and trapping (Pull)
The three component blend of *exo*-brevicomin, frontalin, and myrcene is an extremely effective lure used in multiple funnel or panel traps for monitoring WPB populations (Bedard et al. 1980, Wood 1972). While a large trap-out study using this pheromone blend suggested that the technique may have promise for control of WPB, further wide-scale testing has not been conducted. The recent advances made in finding effective anti-aggregation semiochemicals (Erbilgin et al. 2008, Fettig et al. 2008a, 2008b), however, may reinvigorate this line of investigation as part of a push-pull strategy.

Push
An operational anti-aggregation method for the WPB is not presently available except for single-tree treatments (Fettig et al. 2008a), but research is active in this area and includes developmental testing of alternative active ingredients and tests of acetophenone and ipsdienol in broadcast dispenser applications for stand-level treatments (Gillette et al. 2009a, 2009b). Active ingredients such as those identified by Fettig et al. (2008b) warrant testing for area-wide stand protection as well as individual tree protection.

Red Turpentine Beetle (RTB)

RTB is normally considered a secondary pest of all pine species (Furniss and Carolin 1977), but recent outbreaks have been reported where RTB acts as a primary tree killer (Rappaport et al. 2001). The introduction of RTB into China has raised concerns about its spread across the entire Holarctic region from Asia into Europe and North Africa, since it appears to attack all species within the genus *Pinus* L., and there is a corridor of pines westward from Asia to Europe (Erbilgin et al. 2007a). In Asia, consequently, there has been a concerted effort to control RTB populations and minimize the spread of this invasive species (Yan et al. 2005). In North America there has been less emphasis on control of RTB than in China, but drought stress is known to exacerbate RTB damage (Smith 1961), leading to concerns that warming climates will result in increased damage and a need for control measures.

Monitoring and trapping (Pull)
The standard commercial lure for RTB has been the three-component blend of α- and β-pinene, and Δ^3-carene in a 1:1:1 ratio (Contech International, formerly Pherotech International) (Hobson et al. 1993). Recently, however, it was shown that Δ^3-carene is the most attractive of these monoterpenes over the range of RTB in both North America and Asia (Erbilgin et al. 2007a), and Δ^3-carene alone is a more effective lure for RTB than the blend in most cases. Although trap-out programs have not been conducted in North America, a regional trap-out program conducted in China, where RTB was accidentally introduced in the mid-1980s, was credited in part with a large reduction in RTB populations (J.H. Sun, Chinese Academy of Sciences, personal communication). RTB is widely polyphagous, so trapping programs are underway at ports in many pine-growing regions where accidental introduction of RTB is a concern.

Push
Verbenone pouches (along with NHVs) (Fettig et al. 2005, 2008a, 2008b) and verbenone flakes (Gillette et al. 2006) have been shown to provide significant protection of individual pines from attack by RTB. The application of verbenone-releasing flakes at the rate of 3.57 oz (15 g) of flakes/tree reduced attack rate by RTB on individual trees to nearly zero compared to control trees (Gillette et al. 2006), so this method gives very good individual tree protection. The application of verbenone-releasing flakes may be warranted in campgrounds, ski resorts, and administrative sites to protect individual trees from attack by red turpentine beetle.

Conclusions

Research and development of semiochemicals for bark beetle control has yielded many products and strategies that have recently come to fruition and are now being used to protect high-value stands on public and private lands. Recent developments with products for aerial application have provided tools that are appropriate over larger areas and sites that are inaccessible for hand-applied release devices. This is an active area of research, and new products—both active ingredients and new release devices--are constantly emerging for reducing bark beetle-caused tree mortality. It is therefore important to stay current with new developments and to contact extension entomologists and pheromone company representatives for the latest available information, as the field is rapidly and constantly changing. We wish to emphasize, however, that the use of semiochemicals to protect forest stands from bark beetle attack is really only a short-term solution to a long-term problem. While semiochemicals can provide significant protection over the short term, long-term vegetation management strategies are required to reduce susceptibility to bark beetle damage (Negrón et al. 2008). The need for semiochemical strategies can be significantly diminished by manipulating age class structure, encouraging species diversity and maintaining lower tree densities. In the face of possible climate shifts, however, there may well be increasing need for semiochemicals to protect high-value areas until vegetation management can be implemented to reduce bark beetle risk. These methods may furthermore be helpful in protecting stands or individual trees during periods of temporary vulnerability such as the periods following wildfire, avalanches, and windstorms. They can also be used as part of an intensive management program that incorporates baited sacrificial trees to temporarily reduce bark beetle risk in climate-stressed stands.

Acknowledgments

We thank John Lundquist (USDA Forest Service, Anchorage, AK) for organizing the symposium where this report was first presented, and Jane Hayes (USDA Forest Service, La Grande, OR) and John Lundquist for preparing this proceedings report. We also thank C.J. Fettig (USDA Forest Service, Davis, CA), K. Gibson (USDA Forest Service, Missoula, MT), R. Kelsey (USDA Forest Service, Corvallis, OR), J. Lundquist, and S. Seybold (USDA Forest Service, Davis, CA), for constructive reviews of the manuscript.

Note: Mention of a product does not constitute recommendation for its use by the USDA Forest Service or the authors.

Literature Cited

Amman, G.D.; Logan, J.A. 1998. Silvicultural control of mountain pine beetle: prescriptions and the influence of microclimate. American Entomologist. 44: 166–177.

Amman, G.D.; Thier, R.W.; Weatherby, J.C.; Rasmussen, L.A.; Munson, A.S. 1991. Optimum dosage of verbenone to reduce infestation of mountain pine beetle [*Dendroctonus ponderosae*] in lodgepole pine [*Pinus contorta* var. *latifolia*] stands of central Idaho. Res. Pap. INT-RP-446. Ogden, UT: U.S. Department of Agriculture, Forest Service, Intermountain Research Station. 6 p.

Andrews, S.L.; Perkins, J.P.; Thrailkill, J.A.; Poage, N.J.; Tappeiner, J.C., II. 2005. Silvicultural approaches to develop northern spotted owl nesting sites, central Coast ranges, Oregon. Western Journal of Applied Forestry. 20: 13–27.

Bedard, W.D.; Tilden, P.E.; Wood, D.L.; Silverstein, R.M.; Brownlee, R.G.; Rodin, J.O. 1969. Western pine beetle: response to its sex pheromone and a synergistic host terpene, myrcene. Science. 164: 1284–1285.

Bedard, W.D.; Silverstein, R.M.; Wood, D.L. 1970. Bark beetle pheromones. Science. 167(3925): 1638–1639.

Bedard, W.D.; Tilden, P.E.; Lindahl, K.Q.J.; Wood, D.L.; Rauch, P.A. 1980. Effects of verbenone and *trans*-verbenol on the response of *Dendroctonus brevicomis* to natural and synthetic attractant in the field. Journal of Chemical Ecology. 6: 997–1014.

Bedard, W.D.; Wood, D. L. 1981. Suppression of *Dendroctonus brevicomis* by using a mass-trapping tactic. In: Mitchell, E. R., ed. Management of insect pests with semiochemicals. New York: Plenum Press:103–114.

Bentz, B.J.; Kegley, S.; Gibson, K.; Thier, R. 2005. A test of high-dose verbenone for stand-level protection of lodgepole and whitebark pine from mountain pine beetle (Coleoptera: Curculionidae: Scolytinae) attacks. Journal of Economic Entomology. 98: 1614–1621.

Bentz, B.J.; Munson, A.S. 2000. Spruce beetle population suppression in northern Utah. Western Journal of Applied Forestry. 15(3):122–128.

Blackford, D.C. 2007. Aspen Grove Trailhead Area, Pleasant Grove, Ranger District, Uinta National Forest. Functional Assistance Report. OFO-TR-07-02. Ogden, UT: US Department of Agriculture, Forest Service, Forest Health Protection. 7 p.

Borden, J.H. 1997. Disruption of semiochemical-mediated aggregation in bark beetles. In: Cardé, R.T.; Minks, A.K., eds. Insect pheromone research: new directions. New York: Chapman and Hall: 421–438.

Borden, J.H.; Birmingham, A.L.; Burleigh, J.S. 2006. Evaluation of the push-pull tactic against the mountain pine beetle using verbenone and non-host volatiles in combination with pheromone-baited trees. Forestry Chronicles. 82: 579–590.

Borden, J.H.; Chong, L.J.; Earle, T.J.; Huber, D.P.W. 2003. Protection of lodgepole pine from attack by the mountain pine beetle, *Dendroctonus ponderosae* (Coleoptera: Scolytidae) using high doses of verbenone in combination with nonhost bark volatiles. Forestry Chronicles. 79: 685–691.

Borden, J.H.; Chong, L.J.; Lindgren, B.S.; Begin, E.J.; Ebata, T.M.; MacLauchlan, L.E.; Hodgkinson, R.S. 1993. A simplified tree bait for the mountain pine beetle. Canadian Journal of Forest Research. 23(6): 1108–1113.

Borden, J.H.; Gries, G.; Chong, L.J.; Werner, R.A.; Holsten, E.H.; Wieser, H.; Dixon, E.A.; Cerezke, H.F. 1996. Regionally-specific bioactivity of two new pheromones for *Dendroctonus rufipennis* (Kirby) (Col., Scolytidae). Journal of Applied Entomology. 120: 321–326.

Borden, J.H.; Lacey, T.E. 1985. Semiochemical-based manipulation of the mountain pine beetle, *Dendroctonus ponderosae* Hopkins: a component of lodgepole pine silviculture in the Merritt Timber Supply area of British Columbia. Zeitschrift fur Angewandte Entomologie. 99(2): 139–145.

Borden, J.H.; Pureswaran, D.S.; Poirier, L.M. 2004. Evaluation of two repellent semiochemicals for disruption of attack by the mountain pine beetle, *Dendroctonus ponderosae* Hopkins (Coleoptera: Scolytidae). Journal of the Entomological Society of British Columbia. 101: 117–123.

Borden, J.H.; Sparrow, G.R.; Gervan, N.L. 2007. Operational success of verbenone against the mountain pine beetle in a rural community. Arboriculture and Urban Forestry 33(5): 318–324.

Browne, L.E.; Wood, D.L.; Bedard, W.D.; Silverstein, R.M.; West, J.R. 1979. Quantitative estimates of the western pine beetle attractive pheromone components, *exo*-brevicomin, frontalin, and myrcene in nature. Journal of Chemical Ecology. 5(3): 397–414.

Campbell, S.A.; Borden, J.H. 2006. Integration of visual and olfactory cues of hosts and nonhosts by three bark beetle species (Coleoptera: Scolytidae). Ecological Entomology. 31(5): 437–449.

Carroll, A.L.; Taylor, S.W.; Régnière J.; Safranyik. L. 2003. Effects of climate change on range expansion by the mountain pine beetle in British Columbia. In: Shore; T.L.; Brooks, J.E.; Stone, J.E., eds. Mountain pine beetle symposium, challenges and solutions. October 30–31, Kelowna, BC. Victoria, BC: Natural Resources Canada, Information Report BC-X-399. Canadian Forest Service, Pacific Forestry Centre: 223–232.

Clarke, S.R.; Salom, S.M.; Billings, R.F.; Berisford, C.W.; Upton, W.W.; McClellan, Q.C.; Dalusky, M.J. 1999. A scentsible approach to controlling southern pine beetles: two new tactics using verbenone. Journal of Forestry. 97: 26–31.

Conn, J.E.; Borden, J.H.; Scott, B.E.; Friskie, L.M.; Pierce, H.D. Jr.;Oehlschlager, A.C. 1983. Semiochemicals for the mountain pine beetle, *Dendroctonus ponderosae* (Coleoptera: Scolytidae) in British Columbia: field trapping studies. Canadian Journal of Forest Research. 13(2): 320–324.

Cook, S.M.; Khan, Z.R.; Pickett, J.A. 2007. The use of push-pull strategies in integrated pest management. Annual Review of Entomology. 52: 375-400.

DeGomez, T.E.; Hayes, C.J.; Anhold, J.A.; McMillin J.D.; Clancy, K.M.; Bosu, P.P. 2006. Evaluation of insecticides for protecting southwestern ponderosa pines from attack by engraver beetles (Coleoptera: Curculionidae: Scolytinae). Journal of Economic Entomology. 99: 393–400.

DeMars, C.J.; Slaughter, G.W.; Bedard, W.D.; Norick, N.X.; Roettgering, B. 1980. Estimating western pine beetle-caused tree mortality for evaluating an attractive pheromone treatment. Journal of Chemical Ecology. 6(5): 853–866.

Ebata, T. 2004. Current status of mountain pine beetle in British Columbia. Information Report BC-X-399. Victoria, BC: Natural Resources Canada, Canadian Forest Service, Pacific Forestry Centre: 52–56.

Erbilgin, N.; Mori, S.R.; Sun, J.H.; Stein, J.D.; Owen, D.R.; Campos Bolaños, R.; Merrill, L.D.; Raffa, K.F.; Méndez Montiel, T.; Wood, D.L.; Gillette, N.E. 2007a. Response to host volatiles by native and introduced populations of *Dendroctonus valens* (Coleoptera: Curculionidae, Scolytinae) in North America and China. Journal of Chemical Ecology. 33: 131–146.

Erbilgin, N.; Gillette, N.E.; Mori, S.R.; Stein, J.D.; Owen, D.R.; Wood, D.L. 2007b. Acetophenone as an anti-attractant for the western pine beetle, *Dendroctonus brevicomis* LeConte (Coleoptera: Scolytidae). Journal of Chemical Ecology. 33: 817–823.

Erbilgin, N.; Gillette, N.E.; Owen, D.R.; Mori, S.R.; Nelson, A.S.; Uzoh, F.; Wood, D.L. 2008. Acetophenone superior to verbenone for reducing attraction of western pine beetle *Dendroctonus brevicomis* to its aggregation pheromone. Agricultural and Forest Entomology. 10(4): 433–441.

Fettig, C.J.; McKelvey, S.R.; Huber, D.P.W. 2005. Nonhost angiosperm volatiles and verbenone disrupt response of western pine beetle, *Dendroctonus brevicomis* (Coleoptera: Scolytidae), to attractant-baited traps. Journal of Economic Entomology. 98: 2041-2048.

Fettig, C.J.; Allen, K.K.; Borys, R.R.; Christopherson, J.; Dabney, C.P.; Eager, T.J.; Gibson, K.E.; Hebertson, E.G.; Long, D.F.; Munson, A.S.; Shea, P.J.; Smith, S.L., Haverty, M.I. 2006a. Effectiveness of bifenthrin (Onyx) and carbaryl (Sevin SL) for protecting individual, high-value conifers from bark beetle attack (Coleoptera: Curculionidae: Scolytinae) in the Western United States. Journal of Economic Entomology. 99: 1691–1698.

Fettig, C.J.; DeGomez, T.; Gibson, K.E.; Dabney, C.P.; Borys, R.R. 2006b. Effectiveness of permethrin plus-C (MasterlineReg.) and carbaryl (Sevin SLReg.) for protecting individual, high-value pines (*Pinus*) from bark beetle attack. Arboriculture and Urban Forestry. 32: 247–252.

Fettig, C.J.; McMillin, J.D.; Anhold, J.A.; Hamud, S.M.; Borys, R.R.; Dabney, C.P.; Seybold, S.J. 2006c. The effects of mechanical fuel reduction treatments on the activity of bark beetles (Coleoptera: Scolytidae) infesting ponderosa pine. Forest Ecology and Management. 230: 55–68.

Fettig, C.J.; Klepzig, K.D.; Billings, R.F.; Munson, A.S.; Nebeker, T.E.; Negron, J.F.; Nowak, J.T. 2007. The effectiveness of vegetation management practices for prevention and control of bark beetle infestations in coniferous forests of the western and southern United States. Forest Ecology and Management. 238: 24–53.

Fettig, C.J.; Dabney, C.P.; McKelvey, S.R.; Huber, D.P.W. 2008a. Nonhost angiosperm volatiles and verbenone protect individual ponderosa pines from attack by western pine beetle and red turpentine beetle (Coleoptera: Curculionidae, Scolytinae). Western Journal of Applied Forestry. 23(1): 40–45.

Fettig, C.J.; McKelvey, S.R.; Dabney, C.P.; Borys, R.R.; Huber, D.P.W. 2008b. Response of *Dendroctonus brevicomis* to different release rates of nonhost angiosperm volatiles and verbenone in trapping and tree protection studies. Journal of Applied Entomology. doi:10.1111/j.1439-0418.2008.01317.x

Furniss, M.M.; Clausen, R.W.; Markin, G.P.; McGregor, M.D.; Livingston, RL. 1981. Effectiveness of Douglas-fir beetle antiaggregative pheromone applied by helicopter. Gen.Tech. Rep. INT-GTR-10. Ogden, UT: U.S. Department of Agriculture, Forest Service, Intermountain Research Station. 7 p.

Furniss, M.M.; Markin, G.P.; Hager, V.J. 1982. Aerial application of Douglas-fir beetle antiaggregative pheromone: equipment and evaluation. Gen. Tech. Rep. INT-137. Ogden, UT: U.S. Department of Agriculture, Forest Service, Intermountain Research Station. 9 p.

Furniss, R.L.; Carolin, V.M. 1977. Western Forest Insects. Miscellaneous Publication 273. Washington DC: US Department of Agriculture, Forest Service 654 p.

Gibson, K.; Kegley, S. 2004. Testing the efficacy of verbenone in reducing mountain pine beetle attacks in second-growth ponderosa pine. Forest Health Protection Report, Report 04-7. Missoula, MT: U.S. Department of Agriculture, Forest Service, Forest Health Protection, Northern Region: 8 p. http://www.fs.fed.us/r1-r4/spf/fhp/publications/bystate/R1Pub04-7_verbenone_mpb.pdf

Gillette, N.E.; Erbilgin, N.; Webster, J.N.; Pederson, L.; Mori, S.R.; Stein, J.D.; Owen, D.R.; Bischel, K.M.; Wood, D.L. 2009a. Aerially applied verbenone-releasing flakes protect *Pinus contorta* stands from attack by *Dendroctonus ponderosae* in California and Idaho. Forest Ecology and Management. 257: 1405–1412

Gillette, N.E.; Mehmel, C.J.; Erbilgin, N.; Mori, S.R.; Webster, J.N.; Wood, D.L.; Stein, J.D. 2009b. Aerially applied methylcyclohexenone-releasing flakes protect *Pseudotsuga menziesii* stands from attack by *Dendroctonus pseudotsugae.* Forest Ecology and Management. 257(4): 1231–1236

Gillette, N.E.; Stein, J.D.; Owen, D.R.; Webster, J.N.; Fiddler, G.O.; Mori, S.R.; Wood, D.L. 2006. Verbenone-releasing flakes protect individual *Pinus contorta* trees from attack by *Dendroctonus ponderosae* and *Dendroctonus valens* (Coleoptera: Curculionidae, Scolytinae). Agricultural and Forest Entomology. 8: 243–251.

Graves, A.D.; Holsten, E.H.; Ascerno, M.E.; Zogas, K.P.; Hard, J.S.; Huber, D.P.W.; Blanchette, R.A.; Seybold, S.J. 2008. Protection of spruce from colonization by the bark beetle, *Ips perturbatus*, in Alaska. Forest Ecology and Management. 256(11): 1825–1839.

Gray, D.R.; Borden, J.H. 1989. Containment and concentration of mountain pine beetle (Coleoptera: Scolytidae) infestations with semiochemicals: validation by sampling of baited and surrounding zones. Journal of Economic Entomology. 82: 1399–1495.

Haverty, M.I.; Shea, P.J.; Hoffman, J.T.; Wenz, J.M.; Gibson, K.E. 1998. Effectiveness of esfenvalerate, cyfluthrin, and carbaryl in protecting individual lodgepole pines and ponderosa pines from attack by *Dendroctonus* spp. Res. Pap. PSW-RP-237. Berkeley, CA: U.S. Department of Agriculture, Forest Service, Pacific Southwest Research Station. 12 p.

Hicke, J.A.; Jenkins, J.C. 2008. Mapping lodgepole pine stand structure susceptibility to mountain pine beetle attack across the western United States. Forest Ecology and Management. 255: 1536–1547.

Hicke, J.A.; Logan, J.A.; Powell, J.; Ojima, D.S. 2006. Changing temperatures influence suitability for modeled mountain pine beetle (*Dendroctonus ponderosae*) outbreaks in the western United States. Journal of Geophysical Research. 111: [Not paged] G02019, doi:1029/2005JG000101.

Hobson, K.R.; Wood, D.L.; Cool, L.G.; White, P.R.; Ohtsuka, T.; Kubo, I.; Zavarin, E. 1993. Chiral specificity in responses by the bark beetle *Dendroctonus valens* to host kairomones. Journal of Chemical Ecology. 19: 1837–1847.

Holsten, E.H.; Shea, P.J.; Borys, R.R. 2003. MCH released in a novel pheromone dispenser prevents spruce beetle, *Dendroctonus rufipennis* (Coleoptera: Scolytidae) attacks in south-central Alaska. Journal of Economic Entomology. 96: 31–34.

Holsten, E.H.; Webb, W.; Shea, P.J.; Werner, W.A. 2000. Release rates of methylcyclohexenone and verbenone from bubblecap and bead releasers under field conditions suitable for the management of bark beetles in California, Oregon, and Alaska. Res. Pap. PNW-RP-544. Portland, OR: U.S. Department of Agriculture, Forest Service, Pacific Northwest Research Station. 28 p.

Huber, D.P.W.; Borden, J.H. 2001. Protection of lodgepole pines from mass attack by mountain pine beetle, *Dendroctonus ponderosae*, with nonhost angiosperm volatiles and verbenone. Entomologia Experimentalis et Applicata. 99: 131–141.

Kegley, S.; Gibson, K. 2004. Protecting whitebark pine trees from mountain pine beetle attack using verbenone. Forest Health Protection Report 04-8. Missoula, MT: U.S. Department of Agriculture, Forest Service, Forest Health Protection, Northern Region. 4 p.

Kegley, S.; Gibson, K.; Schwandt, J.; Marsden, M. 2003. A test of verbenone to protect individual whitebark pine from mountain pine beetle attack. Forest Health Protection Report. Report 03-9. Missoula, MT: U.S. Department of Agriculture, Forest Service, Forest Health Protection, Northern Region. 6 p.

Kogan, M. 1998. Integrated pest management: historical perspectives and contemporary developments. Annual Review of Entomology. 43: 243–270.

Krist, F.J.; Jr., Sapio, F.J.; Tkacz, B.M. 2007. Mapping risk from forest insects and diseases. FHTET-2007-6. Morgantown, WV: U.S. Department of Agriculture, Forest Service, Forest Health Protection, Forest Health Technology Enterprise Team. 125 p.

Kurz, W.A.; Dymond, C.C.; Stinson, G.; Rampley, G.J.; Neilson, E.T.; Carroll, A.L.; Ebata, T.; Safranyik, L. 2008a. Mountain pine beetle and forest carbon feedback to climate change. Nature. 452: 987–990.

Kurz, W.A.; Stinson, G.; Rampley, G.J.; Dymond, C.C.; Neilson, E.T. 2008b. Risk of natural disturbance makes future contribution of Canada's forests to the global carbon cycle highly uncertain. Proceedings of the National Academy of Sciences. 105(5): 1551–1555.

Lindgren, B.S.; Borden, J.H. 1993. Displacement and aggregation of mountain pine beetles, *Dendroctonus ponderosae* (Coleoptera: Scolytidae) in response to their antiaggregation and aggregation pheromones. Canadian Journal of Forest Research. 23(2): 286–290.

Lister, C.K.; Schmid, J.M.; Mata, S.A.; Haneman, D.; O'Neil, C.; Pasek, J.; Sower, L. 1990. Verbenone bubble caps ineffective as a preventive strategy against mountain pine beetle attacks in ponderosa pine. Res. Note RM-501. Ogden, UT: U.S. Department of Agriculture, Forest Service, Rocky Mountain Forest and Range Experiment Station. 3 p.

Logan, J.A.; Powell, J.A. 2001. Ghost forests, global warming, and the mountain pine beetle (Coleoptera: Scolytidae). American Entomologist. 47: 160–173.

Naumann, K.; Rankin, L.J. 1999. Pre-attack systemic applications of a neem-based insecticide for control of the mountain pine beetle, *Dendroctonus ponderosae* Hopkins (Coleoptera: Scolytidae). Journal of the Entomological Society of British Columbia. 96: 13–19.

Negrón, J.F.; Bentz, B.J.; Fettig, C.J.; Gillette, N.E.; Hansen, E.M.; Hayes, J.L.; Kelsey, R.G.; Lundquist, J.E.; Lynch, A.M.; Progar, R.A.; Seybold, S.J. 2008. US Forest Service bark beetle research in the western United States: Looking toward the future. Journal of Forestry. 106: 325–331.

Negrón, J.F.; Anhold, J.A.; Munson, A.S. 2001. Within-stand spatial distribution of tree mortality caused by the Douglas-fir beetle (Coleoptera: Scolytidae). Environmental Entomology. 30: 215–224.

Nordlund, D.A.; Lewis, W.J. 1981. Semiochemicals: A review of the terminology. In: Nordlund, D.A.; Jones, R.L.; Lewis, W.J., eds. Semiochemicals: their role in pest control. New York: John Wiley & Sons: 13–28.

Ono, H. 2004. The mountain pine beetle: scope of the problem and key issues in Alberta. Information Report BC-X-399. Victoria, BC: Natural Resources Canada, Canadian Forest Service, Pacific Forestry Centre: 62-66.

Poland, T.M.; Borden, J.H.; Stock, A.J.; Chong, L.J. 1998. Green leaf volatiles disrupt responses by the spruce beetle, *Dendroctonus rufipennis*, and the western pine beetle, *Dendroctonus brevicomis* (Coleoptera: Scolytidae) to attractant-baited traps. Journal of the Entomological Society of British Columbia. 95: 17-24.

Progar, R.A. 2003. Verbenone reduces mountain pine beetle attack in lodgepole pine. Western Journal of Applied Forestry. 18: 229–232.

Progar, R.A. 2005. Five-year operational trial of verbenone to deter mountain pine beetle (*Dendroctonus ponderosae*; Coleoptera: Scolytidae) attack of lodgepole pine (*Pinus contorta*). Environmental Entomology. 34: 1402–1407.

Pureswaran D.S.; Borden, J.H. 2005. Primary attraction and kairomonal host discrimination in three species of *Dendroctonus* (Coleoptera: Scolytidae). Agricultural and Forest Entomology. 7(3): 219-230.

Rappaport, N.G.; Owen, D.R.; Stein, J.D. 2001. Interruption of semiochemical-mediated attraction of *Dendroctonus valens* (Coleoptera: Scolytidae) and selected nontarget insects by verbenone. Environmental Entomology. 30: 837–841.

Ross, D.W.; Daterman, G.E. 1994. Reduction of Douglas-fir beetle infestation of high-risk stands by antiaggregation and aggregation pheromones. Canadian Journal of Forest Research. 24: 2184–2190.

Ross, D.W.; Daterman, G.E. 1998. Pheromone-baited traps for *Dendroctonus pseudotsugae* (Coleoptera: Scolytidae): influence of selected release rates and trap designs. Journal of Economic Entomology. 91: 500–506.

Ross, D.W.; Daterman, G.E.; Munson, A.S. 1996. Optimal dose of an antiaggregation pheromone (3-methylcyclohex-2-en-1-one) for protecting live Douglas-fir from attack by *Dendroctonus pseudotsugae* (Coleoptera: Scolytidae). Journal of Economic Entomology. 89: 1204–1207.

Ross, D.W.; Daterman, G.E.; Munson, A.S. 2002. Elution rate and spacing of antiaggregation pheromone dispensers for protecting live trees from *Dendroctonus pseudotsugae* (Coleoptera: Scolytidae). Journal of Economic Entomology. 95: 778–781.

Ross, D.W.; Daterman, G.E.; Munson, A.S. 2004. Evaluation of the antiaggregation pheromone, 3-methylcyclohex-2-en-1-one (MCH), to protect live spruce from spruce beetle (Coleoptera: Scolytidae) infestation in southern Utah. Journal of the Entomological Society of British Columbia. 101: 145–146.

Ross, D.W.; Daterman, G.E.; Munson, A.S. 2005. Spruce beetle (Coleoptera: Scolytidae) response to traps baited with selected semiochemicals in Utah. Western North American Naturalist. 65: 123–126.

Ross, D.W.; Hostetler, B.B.; Johansen, J. 2006. Douglas-fir beetle response to artificial creation of down wood in the Oregon Coast Range. Western Journal of Applied Forestry. 21: 117–122.

Salom, S.M.; Grosman, D.M.; McClellan, Q.C.; Payne, T.L. 1995. Effect of an inhibitor-based suppression tactic on abundance and distribution of the southern pine beetle (Coleoptera: Scolytidae) and its natural enemies. Journal of Economic Entomology. 88: 1703–1716.

Seybold, S.J.; Downing, M. 2009. What risk do invasive bark beetles and woodborers pose to forests of the western U.S? A case study of the Mediterranean pine engraver, *Orthotomicus erosus*. In: Hayes, J.L.; Lundquist, J.E., comps. Western Bark Beetle Research Group—a unique collaboration with Forest Health Protection symposium, Society of American Foresters Conference, 23–28 October 2007, Portland, OR. Gen. Tech. Rep. PNW-GTR-784, Portland, OR: U.S. Department of Agriculture, Forest Service, Pacific Northwest Research Station: 111–134.

Seybold, S.J.; Huber, D.P.W.; Lee, J.C.; Graves, A.D.; Bohlmann, J. 2006. Pine monoterpenes and pine bark beetles: a marriage of convenience for defense and chemical communication. Phytochemistry Reviews. 5: 143–178.

Sharov, A.A.; Leonard, D.; Liebhold, A.M.; Roberts, E.A.; Dickerson, W. 2002. "Slow the spread:" a national program to contain the gypsy moth. Journal of Forestry. 100: 30–35.

Shore, T.L.; Safranyik, L.; Lemieux, J.P. 2000. Susceptibility of lodgepole pine stands to the mountain pine beetle: testing of a rating system. Canadian Journal of Forest Research. 30: 44–49.

Silverstein, R.M.; Brownlee, R.G.; Bellas, T.E.; Wood, D.L.; Browne, L.E. 1968. Brevicomin: principal sex attractant in the frass of the female western pine beetle. Science. 158(3817): 889–891.

Silverstein, R.M.; Rodin, J.O.; Wood, D.L. 1966. Sex attractants in frass produced by male *Ips confusus* in ponderosa pine. Science. 154: 509–510.

Skillen, E.L.; Berisford, C.W.; Camann, M.A.; Reardon, R.C. 1997. Semiochemicals of forest and shade tree insects in North America,and management implications. FHTET-96-15. Morgantown,WV: U.S. Department of Agriculture, Forest Service, Forest Health Technology Enterprise Team. 189 p.

Smith, R.F. 1962. Principles of integrated pest control. Proceedings of the North Central Branch, Entomological Society of America. Lanham, MD,17, 7 p.

Smith, R.H. 1961. Red turpentine beetle. Forest Pest Leaflet 55. Washington, DC: U.S. Department of Agriculture, Forest Service. 8 p.

Smith, R.H. 1986. Trapping western pine beetles with baited toxic trees. Res. Note. PSW-RN-382. Berkeley, CA: US Department of Agriculture, Forest Service, Pacific Southwest Forest and Range Experiment Station. 9 p.

Tilden, P.E.; Bedard, W.D.; Wood, D.L.; Stubbs, H.A. 1981. Interruption of response of *Dendroctonus brevicomis* to its attractive pheromone by components of the pheromone. Journal of Chemical Ecology. 7: 183–196.

Vandygriff, J.C.; Rasmussen, L.A.; Rineholt, J.F. 2000. A novel approach to managing fuelwood harvest using bark beetle pheromones. Western Journal of Applied Forestry. 15(4): 183–188.

Werner, R.A.; Hard, J.; Holsten, E.H. 1988. The development of management strategies to reduce the impact of the spruce beetle in south-central Alaska. Northwest Environmental Journal. 4(2): 319–358.

Westfall, J. 2007. 2006 Summary of forest health conditions in British Columbia. Pest Management Report Number 15, Victoria, BC: British Columbia Ministry of Forests and Range. 73 p.

Whitehead, R.J.; Russo, G.L. 2005. 'Beetle-proofed' lodgepole pine stands in interior British Columbia have less damage from mountain pine beetle. Information Report BC-X-402. Victoria, BC: Natural Resources Canada, Canadian Forest Service, Pacific Forestry Centre. 17 p.

Wilent, S. 2005. Mountain pine beetles threaten Canadian, US forests. The Forestry Source. http://www.safnet.org/archive/0505_beetle.cfm (28 February 2006).

Wood, D.L. 1972. Selection and colonization of ponderosa pine by bark beetles. In: van Emden, H.F., ed. Insect/plant relationships. Oxford: Blackwell Science Publications: 101–117.

Wood, D.L.; Browne, L.E.; Ewing, B.; Lindahl, K.; Bedard, W.D.; Tilden, P.E.; Mori, K.; Pitman, G.B.; Hughes, P.R. 1976. Western pine beetle: specificity among enantiomers of male and female components of an attractant pheromone. Science. 192: 896–898.

Wood, D.L.; Stark, R.W.; Waters, W. W.; Bedard, W.D.; Cobb, F.W., Jr. 1985. Treatment tactics and strategies. In: Waters, W.W; Stark, R.W.; Wood, D.L., eds. Integrated pest management in pine-bark beetle ecosystems. New York, New York: John Wiley and Sons:121–140.

Yan, Z.L.; Sun, J.H.; Owen, D.R.; Zhang. Z.N. 2005. The red turpentine beetle, *Dendroctonus valens* LeConte (Scolytidae): an exotic invasive pest of pine in China. Biodiversity and Conservation. 14(7): 1735–1760.

What Risks Do Invasive Bark Beetles and Woodborers Pose to Forests of the Western United States? A Case Study of the Mediterranean Pine Engraver, *Orthotomicus erosus*[1]

Steven J. Seybold and Marla Downing[2]

Abstract

Recently reported, and likely to threaten the health of standing trees in the urban and peri-urban forests of the West, are at least five new subcortical insect/pathogen complexes [*Agrilus coxalis* Waterhouse (Buprestidae) and four species of Scolytidae: *Orthotomicus* (*Ips*) *erosus* (Wollaston), *Hylurgus lignipderda* F., *Scolytus schevyrewi* Semenov, and *Pityophthorus juglandis* Blackman, which vectors the invasive fungus, *Geosmithia* sp.]. Through the Forest Insect and Disease Leaflet and Pest Alert series and other extension-type publications, personnel from USDA Forest Service Research and Development (R&D) have worked closely with USDA Forest Service Forest Health Protection (FHP) specialists in the western regions to disseminate information to the public on the distribution, identification, biology, and potential impact of these new pests to western U.S. forests. Because the Mediterranean pine engraver, *O. erosus*, has the most potential to have a strong impact on conifers in western U.S. forests and elsewhere in North America, we focus on this species as a case study for the development of a species-specific national risk map (=Potential Susceptibility map) to illustrate how USDA Forest Service R&D and USDA Forest Service FHP [in this case the Forest Health Technology Enterprise Team (FHTET)], can work cooperatively to address an issue of pressing national concern.

Keywords: Aleppo pine, bark beetles, invasive insect species, Italian stone pine, risk-mapping.

---

[1] This manuscript was prepared at the request of the compilers to address the WBBRG priority area: Develop methods and strategies for detecting, monitoring, and eradicating or mitigating invasive bark beetles and woodboring insects, which was not included in the 2007 SAF Symposium because of time constraints.

[2] **Steven J. Seybold** is a Research Entomologist, USDA Forest Service, Pacific Southwest Research Station, Chemical Ecology of Forest Insects, 720 Olive Drive, Suite D, Davis, CA 95616; email: sjseybold@gmail.com. **Marla Downing** is a Biological Scientist, USDA Forest Service, Forest Health Technology Enterprise Team, 2150 Centre Ave., Bldg. A, Suite 331, Fort Collins, CO 80526-1891; email: mdowning@fs.fed.us.

Introduction

Native bark and ambrosia beetles (Coleoptera: Scolytidae, *sensu* Wood, 2007 and Platypodidae) and woodborers (broadly defined as Coleoptera: Anobiidae, Bostrichidae, Buprestidae, Cerambycidae, Curculionidae, Lyctidae, Oedemeridae; Hymenoptera: Siricidae; and Lepidoptera: Cossidae and Sesiidae) have historically represented a major threat to forests and wood products of the western U.S. (Furniss and Carolin 1977, Solomon 1995). Because these insect guilds feed on the most vital tissues of trees (phloem, cambium, and sapwood of the main stem, root, and root crown), they are considered to have the highest impact on host growth and reproduction, and thus, have been ranked as the most damaging among all forest insects (Mattson 1988). The impact of these endophytic insects is magnified further by their interactions with fungi (Goheen and Hansen 1993, Paine et al. 1997). With the evolution of multiple native complexes of tree-killing bark beetles (e.g., *Dendroctonus*, *Ips*, and *Scolytus* spp.) and, in rare cases, woodborers (e.g., *Melanophila californica* Van Dyke), these feeding groups of insects have reached the pinnacle of their impact in the drought- (Koch et al. 2007), fire- (Parker et al. 2006), and wind-challenged (Gandhi et al. 2007) coniferous forests of the western U.S.

Throughout much of the development of forest entomology in the West, these coniferous forests have been largely unchallenged by invasive insect species in these guilds. In western U.S. forests, Furniss and Carolin (1977) listed only two bark beetles [*Scolytus multistriatus* (Marsham) and *S. rugulosus* (Müller)], one curculionid stem borer [*Cryptorynchus lapathi* (L.)], and one cerambycid stemborer (*Saperda populnea* L.) as introductions from other continents. None of these insects feeds on conifers, and *C. lapathi* is now considered to be a native holarctic species (D.W. Langor, Canadian Forestry Service, personal communication). However, since the monograph by Furniss and Carolin, increasing numbers of invasive bark beetles and woodborers have been detected and have established populations in urban and wildland forests of the West (Haack 2006; Langor et al. 2008; Lee et al. 2005, 2006, 2007; Liu et al. 2007; Mattson et al. 1992; Moser et al. 2005) (Table 1). Notably, some of these additions to our subcortical forest insect fauna are well-documented pests of conifers on other continents (Table 2).

In this paper we briefly discuss the concept of new invasive subcortical insects in western U.S. forests from the perspectives of: (1) the resources threatened and (2) the risks posed by the invaders. We use the Mediterranean pine engraver, *Orthotomicus (Ips) erosus* (Wollaston), as a case study for the development of a species-specific national risk map to illustrate how USDA Forest Service Research and Development (R&D) and USDA Forest Service Forest Health Protection (FHP) [in this case the Forest Health Technology Enterprise Team (FHTET)] can work cooperatively to address an issue of pressing national concern.

Table 1—Invasive bark and woodboring beetles first detected in the western U.S. between 1984 and 2008[1]

Species	Family	State where initially detected
Heterobostrychus brunneus (Murray)	Bostrichidae	California
Sinoxylon ceratoniae (L.)	Bostrichidae	California
Agrilus coxalis Waterhouse	Buprestidae	California
Agrilus prionurus Chevrolat	Buprestidae	Texas
Phoracantha recurva Newman	Cerambycidae	California
Phoracantha semipunctata (F.)	Cerambycidae	California
Dendroctonus mexicanus Hopkins	Scolytidae	Arizona
Hylurgus ligniperda F.	Scolytidae	California
Orthotomicus erosus (Wollaston)	Scolytidae	California
Phloeosinus armatus Reitter	Scolytidae	California
Scolytus schevyrewi Semenov	Scolytidae	Colorado
Trypodendron domesticum (L.)	Scolytidae	Washington
Xyleborinus alni (Niisima)	Scolytidae	Washington
Xyleborus similis Ferrari	Scolytidae	Texas

[1]We consider Texas to be part of the continental western U.S.; these introductions were documented in Haack (2006), except for *P. semipunctata*, which was reported in Scriven et al. (1986); *D. mexicanus* (Moser et al. 2005); *H. ligniperida* (Liu et al. 2007); *A. coxalis* (Coleman and Seybold in press); and *T. domesticum* (R. Rabaglia, USDA Forest Service, Washington, D.C., personal correspondence).

Recently Introduced Subcortical Insect/Pathogen Complexes in Western U.S. Forests

For a variety of historical, biological, and societal reasons, it appears that the conifer-dominated forests of the western U.S. have accumulated a relatively depauperate fauna of invasive subcortical insects. The situation is similar in Canada where a recent survey of all non-native terrestrial arthropods associated with woody plants revealed that only 12% of these invasive species were bark- and wood-feeders (and this guild was liberally defined to include external feeders on roots and gall makers on twigs) (Langor et al. 2008). Among all of the families of subcortical insects noted above, only one invasive cerambycid, nine invasive scolytids, and one invasive sesiid were listed for western Canada (provinces west of Manitoba). Factors such as species composition, abundance, and locations of native and adventive stands of trees and shrubs; diversity and abundance of competing native species of subcortical insects; historical patterns of trade and land use; historical working locations of collectors and survey entomologists; and locations of urban centers relative to forest lands may all have played a role in the relatively low number of subcortical insects recorded from western North American forests. Forests of the western U.S. have high levels of native biodiversity of conifers as well as bark beetles and woodborers (Bright and Stark 1973, Furniss and Johnson 2002, Little 1971, Wood 1982). Thus, although invading species have had a range of potential hosts at their disposal, they may also have faced greater competition for various niches by native subcortical species. Historically, the contraposition of these factors, and the societal factors listed above, may have made western U.S. forests less vulnerable to invasion by subcortical insects.

Cataloging invasive species is a dynamic process, and the lists developed in the literature are ephemeral (Langor et al. 2008). Nonetheless, of 25 bark and woodboring Buprestidae, Cerambycidae, and Scolytidae first reported to be established in the continental U.S. between 1985 and 2005, seven species were in the western U.S. (Haack 2006). Two other invasive species of woodboring Coleoptera (Bostrichidae), traditionally more associated with wood products, have also been reported from California (Table 1). Established populations of eucalyptus longhorned borer, *Phoracantha semipunctata* (F.) (Cerambycidae), were first discovered in southern California in 1984 (Scriven et al. 1986), but this species was not included in the survey by Haack (2006). In total, established western U.S. populations of at least 14 subcortical insect taxa have been reported in the literature since 1984 from Arizona, California, Colorado, Texas, and Washington (Table 1). Not all of these bark and woodboring taxa are likely to assume pest status in U.S. forests.

However, some of the more recently reported subcortical insect/pathogen complexes are likely to threaten the health of standing trees in the urban, peri-urban, and wildland forests of the West (Table 2).

Table 2—Emerging threats posed by recently detected invasive bark beetles, woodborers, and/or pathogens in the western U.S

Species	Hosts	Fungal associates in U.S. population	Observed levels of tree mortality in the western U.S.	References
Agrilus coxalis[1]	Quercus spp.	Unknown	Locally extensive, wildland urban interface (S. CA)	Coleman and Seybold, 2008a,b
Dendroctonus mexicanus	Pinus spp	Unknown[2]	Locally extensive in a species complex of other Dendroctonus (S. Az)	Moser et al. 2005
Hylurgus ligniperda	Pinus spp.	Ophiostoma ips, O. galeiforme, and ten other ophiostomoid fungi	None	Lee et al. 2007, Liu et al. 2007, 2008, S. Kim and T.C. Harrington personal communication
Orthotomicus erosus	Pinus spp.	Ophiostoma ips	Minor levels, urban forests (CA)	Lee et al. 2005, 2007, 2008, T.C. Harrington personal communication
Pityophthorus juglandis	Juglans spp.	Geosmithia sp.	Westwide, urban forests, rural landscapes (CA, CO, UT)	N.A. Tisserat, personal communication
Scolytus schevyrewi	Ulmus spp.	Ophiostoma novo-ulmi	Locally extensive, urban forests (WY, CO)	Negrón et al. 2005; Jacobi et al. 2007; Johnson et al. 2008 ; Lee et al. 2006, 2007, In press

[1](Coleoptera: Buprestidae); all other species in this table are (Coleoptera: Scolytidae).
[2]Fungal isolations from the U.S. population of *D. mexicanus* were in progress as of Nov. 2008 (K.D. Klepzig, USDA Forest Service, Asheville, NC, and D.L. Six, University of Montana, Missoula, MT, personal communication).

Two of these complexes are on pines in California [*O. erosus* and the redhaired pine bark beetle, *Hylurgus lignipderda* F. (both Scolytidae)]; one is on pines in Arizona [the Mexican pine beetle, *Dendroctonus mexicanus* Hopkins (Scolytidae)]; one is on oaks in California [the goldspotted oak borer, *Agrilus coxalis* Waterhouse (Buprestidae)], and two are on other hardwoods across the West [the banded elm bark beetle, *Scolytus schevyrewi* Semenov, and the walnut twig beetle/thousand cankers complex, *Pityophthorus juglandis* Blackman (Scolytidae) and *Geosmithia* sp.]. *Agrilus coxalis*, *D. mexicanus*, and *P. juglandis* are not invasive insects from other continents, but the recent discoveries of *A. coxalis* and *D. mexicanus* in the U.S. appear to be range expansions or regional introductions (Coleman and Seybold 2008a, Moser et al. 2005); *P. juglandis* appears to be damaging native and adventive stands of walnut trees through an association with an invasive fungal pathogen (N.A. Tisserat, Colorado State University, personal communication). The occurrences of regional introductions or range expansions leading to "indigenous exotic species" may reflect either more lax intracontinental and interstate commercial regulatory enforcement (Dodds et al. 2004) or effects of climate change (Hicke et al. 2006) on native subcortical insect distributions. These subtly continuous or discrete geographical shifts in subcortical forest insect populations may be a challenging wave of the future in invasive species management.

Through the Forest Insect and Disease Leaflet and Pest Alert series and other extension-type publications, personnel from USDA Forest Service R&D have worked closely with specialists from the western regions of USDA Forest Service FHP in conjunction with University of California at Davis entomologists to disseminate information to the public on the distributions, identification, biology, and potential impacts of the new subcortical insect pests to western U.S. forests (Coleman and Seybold 2008; Lee et al. 2005–2007; Liu et al. 2007; Negrón et al. 2005). It appears that most of the invasive species in this ensemble of subcortical insects successfully colonize trees under some form of stress. However, based on the damage that it has caused to stressed pines in other continents, *O. erosus* has perhaps the most potential to have a strong impact on conifers in western U.S. forests and elsewhere in North America.

Orthotomicus erosus: Introduction, Establishment, Biology, and Behavior
In May 2004, a new exotic bark beetle for North America was discovered in baited flight traps in Fresno, California, during an annual bark beetle and woodborer survey led by Richard L. Penrose of the California Department of Food and Agriculture. This bark beetle was identified as *Orthotomicus erosus* (Wollaston), the Mediterranean pine engraver, a well-documented pest of pines in its native range, which includes the Mediterranean region, the Middle East, Central Asia, and China (Eglitis 2000, Mendel and Halperin 1982, Yin et al. 1984). In July 2007, the widespread occurrence and host range of the pest in China was confirmed by one of us (SJS), through an examination of the holdings of the Chinese Academy of Sciences insect collection in Beijing. How the beetle entered the U.S. is unknown, but it may have arrived with solid wood packing material associated with imported goods. In a survey of records from the USDA APHIS Port Information Network (1985–2001) (Haack 2001), *O. erosus* was the second most frequently intercepted bark beetle species at U.S. ports with a total of 385 interceptions.

Beetles were most frequently associated with imports from the following countries in descending order: Spain, Italy, China, Turkey, and Portugal. Based on remnants of old galleries observed in dead standing trees and in weathered cut pine logs, this beetle was likely present in California for at least three years before its detection in 2004. The distances between the observation points of some of these remnant galleries, the widespread occurrence of *O. erosus* in the state (see below), and its marked abundance, all suggest that this is a minimum estimate of the initial introduction of the species to California.

Since the initial detection, this species has been found in flight traps or has been collected in host material in ten counties in California, primarily in the southern Central Valley (R.L. Penrose, CDFA, unpublished data). Furthermore, in Fresno, Tulare, and Kern Counties, abundant overwintering populations of larvae, pupae, and adults have been found in cut logs of Aleppo, *Pinus halepensis*, Canary Island, *Pinus canariensis*, and Italian stone pine, *Pinus pinea*. These exotic trees are a frequent and esthetically important component of the urban forests of the southern Central Valley and the Los Angeles Basin (Seybold et al. 2006b). They are also widely planted along highway corridors and as shelterbelts in rural regions of California. The Mediterranean pine engraver has so far been detected in urban and peri-urban locations, particularly parks, golf courses, and green waste recycling facilities. The highest population density appears to be in the southeastern Central Valley along the somewhat industrialized State Highway 99 corridor.

Orthotomicus erosus adults generally behave as secondary pests. They are most likely to infest recently fallen trees, standing trees that are under stress, logging debris, and broken branches with rough bark that are at least 5 cm in diameter. Healthy trees have rarely been attacked. In Israel, beetles are often found on the main stem and larger branches of stressed trees that are over 5 yr old (Halperin et al. 1983). In California, this species, or evidence of its past activity, has been found in cut logs from 15 to 90 cm in diameter, on stumps from 10 to 90 cm in diameter, on declining branches of live standing trees, and on the main stem of moribund or dead standing trees. This species has two or more generations per year in its native range. In California, it has three to four generations per year and adults are active year round with the exception of a short period between mid-December and late-January (Lee et al. 2007, In press; R.L. Penrose, CDFA, unpublished data).

Males are the initial sex to colonize cut logs or fallen or standing trees. They construct a nuptial chamber, in which they are typically joined by two females (Mendel and Halperin 1982). Males produce an aggregation pheromone consisting of 2-methyl-3-buten-2-ol and (+)-ipsdienol, whose attraction is enhanced by the host monoterpene α-pinene (Lee et al. In prep., Seybold et al. 2006a). The monoterpene co-attractant is important to the activity of this pheromone in contrast to related beetles in the genus *Ips* where monoterpenes play a relatively minor role (Seybold et al. 2006a). Once in the nuptial chamber, each *O. erosus* female mates and constructs an egg gallery in opposite directions and in longitudinal orientation with the grain of the wood. In the relatively rare (approx. 4%) instances when a third female joins the familial gallery, she excavates her egg gallery parallel to one already established (Mendel and Halperin 1982). Each female lays 26 to 75 eggs and may leave the gallery to lay eggs in a second gallery (Mendel and Halperin 1982). The eggs hatch and larvae develop through three instars expanding the tunnels as they feed. As the tunnels expand, they may overlap with one another. When larvae are ready to pupate, they tunnel towards the bark, especially in cases where the phloem is thick such as *P. canariensis* and *P. pinea*. Observations in the Central Valley of California indicate that this species overwinters as larvae, pupae, and adults beneath the bark surface. Overwintered adult beetles start flying in January and February and establish brood galleries by mid-March. Subsequent broods are initiated in early June, late July, and over an extended period between early September and late November (R.L. Penrose, CDFA, unpublished data). Flight of parent and new adults continues until November and even early December (Lee et al. In press). In Israel, adults start brood production in early March and require a period of feeding before reaching sexual maturation (Mendel 1983). When beetles complete their development, the adults emerge, leaving a small round exit hole in the outer bark, approx. 1.5 mm in diameter. During the warmer parts of the season, there is a bimodal, diurnal pattern of adult flight dispersal with peaks in the morning and evening (Mendel et al. 1991). This has been noted in California as well (D.-G. Liu, SJS, personal observations). New adults may re-infest the same host material that they emerged from or may attack new material.

Laboratory studies in Israel have provided data on the lower temperature thresholds for various aspects of the life history of *O. erosus* (Mendel and Halperin 1982). Females oviposited between 18 and 42°C, but eggs exposed to lower temperatures did not hatch below 16–17°C. Larvae exposed to lower temperatures did not complete development below 18°C (they fed and developed, albeit not completely, for a short period at 14°C). Prepupal development was delayed at temperatures below 16 to 17°C, but individuals did continue to develop at 14°C and became adults after 30 d. In the field in Israel, *O. erosus* developed in areas where winter temperatures ranged from 7.8 to 14°C. As long as the adults initiated the life cycle during periods of warmer temperatures, the immature stages developed through the winter, likely during periods when daily temperatures exceeded 18 to 20°C. The cold temperature tolerances of *O. erosus* have not been studied in the field in California, but lower lethal temperatures and supercooling points of the California population are being investigated in laboratory studies in the Minnesota Department of Agriculture-University of Minnesota BL2 Quarantine Facility in St. Paul, Minnesota (Venette et al. 2009).

In addition to its native range and the recent introduction in California, *O. erosus* has also been introduced into Chile, South Africa, and Swaziland. In all of these locations, the beetle reproduces in a variety of pines, including some that occur in native stands or ornamental plantings in the U.S. (Eglitis 2000). Outside the U.S., *O. erosus* has also been found in Douglas-fir, *Pseudotsuga menziesii* (Mirb.) Franco, spruce, *Picea* sp., fir, *Abies* sp., cypress, *Cupressus* sp., and cedar, *Cedrus* sp., but these non-pine hosts were thought to be used mainly for maturation feeding or overwintering sites for adults (Eglitis 2000). Recently, in laboratory no-choice host range tests of 22 conifers, Lee et al. (2008) reported that *O. erosus* reproduced on four pines from its native Eurasian range—Aleppo, Canary Island, Italian stone, and Scots pines; 11 native North American pines—eastern white, grey, jack, Jeffrey, loblolly, Monterey, ponderosa, red, Sierra lodgepole, singleleaf pinyon, and sugar pines; and four native non-pines— Douglas-fir, black and white spruce, and tamarack. Among non-pines, fewer progeny developed and were of smaller size on Douglas-fir and tamarack, *Larix laricina* (Du Roi) K. Koch, and the number of progeny did not replace the number of founder adults in tamarack. Beetles did not develop on white fir, incense cedar, or coast redwood.

Although *O. erosus* is not a tree-killing bark beetle under normal circumstances, it has demonstrated the capacity to kill trees following disturbances. These have included forest thinning followed by drought in Israel (Halperin et al. 1983; Mendel and Halperin 1982); forest thinning alone in Israel (Mendel et al. 1992); and fire in South Africa (Baylis et al. 1986). Bevan (1984) also provides anecdotal evidence of the reaction of populations of *O. erosus* to various pre-disposing factors in Swaziland; Zwolinski et al. (1995) suggest that in South Africa, *O. erosus* has a higher rate of infestation in pines that were previously wounded by hail and infected with fungi through the wounds. In one instance where *O. erosus* has killed trees in the apparent absence of any pre-disposing conditions, Jiang et al. (1992) reported that *O. erosus* colonized healthy *P. massoniana* and caused a 20% loss of standing pines in the Zhejiang University Forest in China. Seybold and colleagues have observed about 10 cases where *O. erosus* has colonized the main stem of standing ornamental or windbreak pines in the Central Valley of California, but in each instance it was not clear whether the trees were first declining due to some other factor perhaps related to moisture or root pathogens.

Besides direct injury to pine trees, *O. erosus* can vector fungal pathogens. In South Africa, spores of *Ophiostoma ips* (Rumb.) Nannf., the causative agent of bluestain fungus, were found on 60% of 665 adult beetles or galleries on trap logs of *Pinus elliottii* Engelm.and *P. patula* Scheide & Deppe ex Schlech. & Cham.; spores of *Leptographium lundbergii* Lagerb. & Melin were also found on a few samples (Zhou et al. 2001, 2002). Spores of *Graphium pseudormiticum* Mouton & Wingfield have been found with *O. erosus* on unspecified pine logs (Mouton et al. 1994). In Spain a small proportion of a sample population of *O. erosus* were reported to carry the pitch canker fungus, *Fusarium circinatum* Nirenberg and O'Donnell (Romon et al. 2007). In California, the mycoflora of *O. erosus* overwintering in *P. canariensis* and *P. halepensis* was heavily dominated by *Ophiostoma ips* (S. Kim et al. unpublished data, Iowa State University), which agrees with phytopathological studies of *O. erosus* in South Africa (see above) and North Africa (Ben Jamaa et al. 2007).

Given this background, *O. erosus* presents a relatively high risk to pines in North America. It has been included in the ExFor database (http://spfnic.fs.fed.us/exfor) as one of many species of concern to North American forests. Its establishment in much of the U.S. seems highly probable; subsequent spread is likely to cover a large geographic area; and economic damage is likely to be severe (Eglitis 2000). Plantation pines in the southeastern U.S. are particularly vulnerable because both climate and hosts are likely to be favorable. The National Plant Board recognized the threat that *O. erosus* poses for U.S. pines, and USDA APHIS considers the insect as "actionable" (J.F. Cavey, USDA

APHIS, personal correspondence). Out of this regulatory climate, in February 2007 the USDA Forest Service, FHTET with the support of the FHP Early Detection-Rapid Response National Program Coordinators formed an advisory committee chaired by one of us (MD) to develop a U.S. national risk map (Potential Susceptibility Map) for *O. erosus*. Input into the mapping process and research data on the behavior of the invasive population in California was sought from Forest Service R&D, so SJS and Research Biologist R.C. Venette (Northern Research Station) were invited to participate on the committee[1]. What follows is a brief overview of the process, progress, and pitfalls encountered in the development of the *Orthotomicus erosus* Potential Susceptibility Map with an emphasis on the cooperative synergy achieved between R&D and FHP.

Development of the *Orthotomicus erosus* Potential Susceptibility Map

The *Orthotomicus erosus* Potential Susceptibility Map describes the relative potential for introduction and establishment of *O. erosus* at any given location in the conterminous U.S. (USDA FS 2008). The map was constructed by using techniques developed for the 2006 National Insect and Disease Risk Map (Krist et al. 2007), and for susceptibility maps for the invasive subcortical insects, *Ips typographus* (L.) (Coleoptera: Scolytidae) and *Sirex noctilio* F. (Hymenoptera: Siricidae), as well as the invasive pathogen, *Phytophthora alni* Brasier & S.A.Kirk (USDA FS 2008). The techniques rely on a compilation of both expert opinion and research findings, which are synthesized into a series of interacting layers with Geographic Information Systems (GIS) technology (i.e., ESRI 2008, Table 3). Location and level of risk are mapped in 1 km pixels.

The first major component of the map is the Potential Introduction model, which describes where the pest is most likely to enter and escape into the U.S. The model contains information about major ports, markets (=municipalities), and inland distribution centers (Table 3). A dispersal function (Table 4) is used to predict the movement of *O. erosus* from each of these potential points of introduction.

The second major component is the Potential Establishment model, which describes where the pest can survive and reproduce should it be introduced (Table 3). The model contains data related to the temperature tolerances and the host range of *O. erosus*, and a disturbance layer, which is the most data intensive portion of the effort. The factors that go into this layer include ozone, drought (= moisture deficit), fire, hurricane, tornado, avalanche, lightning, and extreme wind events.

---

[1]Other committee members include: D. Borchert, F.H. Koch, F. Krist, F. Sapio, W.D. Smith, S. Smith, B. Tkacz, and M. Tuffly.

Table 3—Data layers included in the *Orthotomicus erosus* Potential Susceptibility Model

Data Layer	Purpose
Introduction: Marine Ports[1]	U.S. marine ports that handle commodities and solid wood packing materials shipped from countries with established populations of *O. erosus*. These are locations where *O. erosus* may be released.
Introduction: Markets	Possible destination locations where *O. erosus* may be released.
Introduction: Inland Distribution Centers	Possible destination locations where *O. erosus* may be released.
Establishment: Temperature Tolerance	A limiting maximum coupled with a minimum temperature range within which *O. erosus* can survive.
Establishment: Host Range	Tbe distribution of tree species, which are used by *O. erosus*, for growth and reproduction.
Establishment: Disturbance	Depicts locations where natural or anthropogenic events occur and potentially affect tree health and vigor; *O. erosus* population densities increase in stressed trees.

[1]The volume of imports into these ports was not considered in the analysis, but the types of imported goods (i.e., those with solid wood packing materials) were taken into consideration. Included in the analysis were all ports were previous USDA Animal and Plant Health Inspection Service interceptions of *O. erosus* had occurred.

**Table 4—Distance-Decay (Dispersal) Function for
the Probable Flight Range of *Orthotomicus erosus***

Distance (km)	GRID Value
0 (Source)	10
GE 1 and LT or EQ to 2	10
GT 2 and LT or EQ to 3	3
GT 3 and LT or EQ to 4	1
GT 4	0

Abbreviations include GE: greater than or equal to;
LT: less than; EQ: equal to; and GT: greater than.

The final Susceptibility Model is a weighted overlay of the Introduction and Establishment components. For every pixel location (i), the values of each spatially coincident Introduction pixel (I_i) and Establishment pixel (E_i) are multiplied by assigned weights (x_I and x_E, respectively), then these values are multiplied, and the product is applied to the Susceptibility pixel (S_i).

$$S_i = I_i (x_I) \cdot E_i (x_E)$$

Without reason to do otherwise, the Susceptibility Map should be the result of an equally-weighted overlay of the Introduction and Establishment components (i.e., $x_I = x_E$ =0.5). Given sufficient reason though, it is possible to attribute more importance to one component by assigning different weights. For example, if the pest is not thought to already have been introduced and the dispersal distance of the pest is relatively limited, it may be more accurate to emphasize the Introduction component. Multiplying the Introduction pixel values by a greater weighting factor and the Establishment pixel values by the complementary factor (before multiplying the two products to create the Susceptibility Map) will prioritize areas where the pest is likely to first be released. On the other hand, if the pest is known to have already been introduced, prioritizing areas where the pest is most able to survive may be desired (i.e., $x_E > x_I$). In this case, the resulting Potential Susceptibility Map will allow pest specialists to focus detection efforts in areas where introduced populations of *O. erosus* may be expanding.

Contributions from Forest Service Research and Development (R&D) to the *Orthotomicus erosus* Potential Susceptibility Map

Research on *O. erosus* in California has provided data for the Introduction and Establishment components on the physiological host range and life history of the insect, the current distribution of the invasive population in California, and the innate flight capacity of adults. The latter is being studied through mark-recapture flight experiments in extremely level and open agricultural fields located in Kings and Tulare Co., California. There are no trees, and specifically no host trees, located within the immediate study areas. Although it is less than half the size of a grain of rice, *O. erosus* is a relatively strong flier that can move at least 10 km in a matter of 24-48 hr with prevailing winds (D.-G. Liu et al., unpublished data).

In addition, Forest Service R&D personnel provided additional locations for a series of inland commercial distribution centers identified during field research and population surveys conducted in the zone of infestation in California. These additional locations were provided to FHTET for incorporation into the Introduction component and as a consequence, similar distribution center data were collected on a national basis and included in the process. R&D personnel also guided the interpretation of the scientific literature for the incorporation of the impact of temperature on developmental thresholds for *O. erosus* into the Establishment component (Mendel 1983, Mendel and Halperin 1982, NAPPFAST 2008). A lower critical development threshold between 0 and 10°C was chosen for the analysis. R&D personnel have also provided advice on weighting various potential hosts and the urban vs. wildland habitats in the Establishment component.

No new data were available from California to aid in the development of the disturbance layer. Likely because it is early in the invasion phase, surveys to date by Forest Service R&D personnel and CDFA cooperators have revealed that populations of the beetle are confined to urban and rural agricultural areas and have not invaded the National Forest system or commercial forest lands in California. Thus, observations of the impact of disturbances such as fire, wind, thinning, etc. could not be recorded. Nevertheless, R&D encouraged strong consideration of the potential interactive power of thinning, drought, and ozone on the health of host pines (Grulke et al. 2002) in the development of this data layer. Through administrative access provided by the USDA FS Southern Research Station, the committee was able to include expertise on drought impacts (Koch et al. 2007) and ozone damage bioindicator data in the modeling procedure.

Finally, research data on the current distribution of *O. erosus* in California and the dispersal function (Table 4) have been combined to test the predictive power of the *O. erosus* Potential Susceptibility Map at locations where the beetle has been flight trapped or hand collected in host material in California.

Contributions from the Forest Health Technology Enterprise Team to the *Orthotomicus erosus* Potential Susceptibility Map

The role of FHTET is to develop technology that assists Forest Health Protection Staff and their cooperators in the management of North America's forests. The specific purpose in developing invasive species tools such as the *O. erosus* Potential Susceptibility model and the resulting map is to provide geographic information for prioritizing detection efforts.

The construction of the *O. erosus* model has required extensive coordination and communication, which have been led by FHTET. Initially, FHTET identified individuals with expertise in risk assessment work, or who had particular knowledge and information about *O. erosus*. Once identified, the participating individuals were informed about the FHTET modeling methods and were invited to participate in the steering committee. Committee members and other experts were then regularly contacted for pertinent knowledge and information on both the biology and behavior of the pest as well as the pest hosts. The knowledge and information from the committee was collected from published research, unpublished documentation, or in the form of personal communication. These inputs were assimilated and essential parameters critical to developing the model were selected (see above).

Once the parameters were selected, an intensive data management effort was undertaken by the FHTET team (i.e., university cooperators and FHTET contractors). Datasets, which were not collected specifically for these purposes, had to be identified and investigated to determine whether they could be used to appropriately characterize the parameters necessary for the model. Often, myriad analyses were required to determine how a dataset can best be utilized to represent the input parameters. Once identified and acquired, representative datasets were processed and standardized and finally combined into Model Builder (ESRI 2008) for inclusion in the model by FHTET contractors.

The process frequently identified knowledge and data gaps, and to address these gaps, multiple versions of the model were provided to the committee. With each new iteration, new issues were discovered and resolved. Resolution of the issues sometimes required that weaknesses in the datasets had to be overcome. This is a difficult issue because it often required the expenditure of a large amount of FHTET resources to re-investigate, analyze, and process the existing data, or to find replacement data. Other issues elucidated the need to: 1) incorporate different and/or additional parameters; 2) set new and/or change parameter thresholds; and 3) make necessary assumptions.

In the development of the *O. erosus* Potential Susceptibility Map, FHTET was responsible for synthesis of information before, during, and after committee meetings; reporting outputs (e.g., loading products on web sites and maintaining web sites); coordination of expertise on the committee (driving the process forward); meeting deadlines; model construction; and hiring contracting and North Carolina State University personnel in order to obtain specific modeling expertise.

Limitations and Pitfalls of the *Orthotomicus erosus* Potential Susceptibility Map
The process of developing a susceptibility map involves a large number of assumptions, such as 1) assigning the magnitude of weights for various factors (see description of this above); 2) developing a course of action when no representative data are available;

3) anticipation of changes in the behavior of *O. erosus* in its new environment; and 4) addressing temporal issues. These assumptions along with all methods and "metadata" for *O. erosus* are disclosed in detail in the "metadata" found on the FHTET website (USDA FS 2008). Some examples of the assumptions follow.

The creation of the urban host layer involved an assumption made to address the lack of available representative data. No comprehensive information exists regarding the presence of host species and the proportion of those species within U.S. urban boundaries. However, it is widely understood that pines are cultivated in nearly all U.S. cities. Therefore, it was assumed that all U.S. cities contain *O. erosus* host type and an urban host layer was created to reflect that decision. In order to capture the maximum number of sample detection points, an expanded definition of urban forest was used that created two levels of risk: 1) the ESRI city (urban) polygons, which introduced a risk level of 7 into the Introduction component; and 2) an urban boundary that begins at the edge of the ESRI polygons and extends outward based on measurements of the lighting footprint created by urban areas (collected from remote sensing imagery), which introduced a risk level of 3-4. An analysis with a dataset assembled by R&D in cooperation with the California Department of Food and Agriculture on the collection locations of *O. erosus* in California revealed that approx. 50% of the detection sample points were in the city lighting areas (= white space).

A number of urban host areas were excluded from the <u>Establishment</u> component because they fell outside the *O. erosus* survival temperature thresholds set by the Committee. Because some of the excluded southwestern urban areas (e.g., Las Vegas, Phoenix, etc.) maintain large ornamental plantings of Mediterranean pines (e.g., *P. canariensis*, *P. eldarica*, *P. halepensis*, and *P. pinea*) susceptible to *O. erosus*, urban areas located within USDA Plant Hardiness Zones of 8b-10b (Cathey 1990) were included in the spatial scope of the model. The Committee concluded that these zones were indicative of the potential of typical Mediterranean pine hosts to grow in these urban areas and that this constituted indirect evidence for potential survival of *O. erosus* in suitable urban microhabitats.

Other assumptions were made to address temporal matters, such as deciding whether more predictive power could be attained with historical datasets (e.g., the preceding 100 yrs) or with contemporary datasets that may reflect present and future conditions (particularly in the context of climate change). Often, the datasets available to the committee were two or three years old because of lags between the time when the data were collected and when they were processed and made available for dissemination. In general though, where relatively current data (e.g., the previous 5 yrs) were available, the committee opted to use these more recent datasets.

The availability and the maintenance of various datasets was also an issue that the committee had to contend with. For example, in the process of developing the disturbance layer, the committee realized that thinning or harvesting was not included as a factor. Given that outbreak activity of this pest on other continents has been correlated with thinning events (Halperin et al. 1983, Mendel and Halperin 1982, Mendel et al. 1992), the absence of this information in the model is a considerable limitation. Unfortunately, the committee ascertained that there was no available national database that consolidates information from National Forests, or state, municipal, private, or Native American lands regarding harvests or thinning operations, so we could not include these data in the map. The committee also found that other datasets in the disturbance layer, e.g., those that document occurrence of tornadoes and hurricanes, were not frequently updated. At the time of this writing (Nov. 2008), the disturbance impact of Hurricane Katrina, which occurred in Aug. 2005, had not been incorporated into the tornado/hurricane database.

The committee also found several instances where the interaction of datasets had to be reconciled. For example, it was accepted by the committee that ozone affects trees that are stressed by drought to a greater extent than trees not experiencing a moisture deficit (Grulke et al. 2002). Therefore, Environmental Protection Agency data on direct ozone concentrations were combined with USDA Forest Service Forest Inventory and Analysis (FIA) program plant damage data, as the latter were thought to better depict the locations where ozone would actually impact the health of pines. Thus, an indirect measure of ozone (i.e., ozone damage to plant bio-indicators) was incorporated into the disturbance layer. Unfortunately, all currently available ozone bioindicator maps had data gaps (i.e., states where bioindicator data were not collected). Although preliminary regional models that relate ozone injury to ambient ozone levels, moisture status, and other environmental variables were available, the committee decided to re-interpolate FIA bioindicator data to fill in the current gaps.

Future Application of the *Orthotomicus erosus* Potential Susceptibility Map
The *O. erosus* Potential Susceptibility Map was developed to understand where *O. erosus* may be entering the U.S. and where it is possible for *O. erosus* to sustain populations. The latter was accomplished by focusing on factors that affect distribution. It was intended that such predicted distributional information could be used by forest resource managers to better direct and pinpoint future monitoring and survey activities for *O. erosus*. In some instances, pest detection specialists may be more interested in using the Potential Introduction portion of the product (i.e., in those states that are far from the current area of infestation); in other instances (e.g., in California and neighboring states) resource managers may be more interested in using the Potential Establishment portion to guide their forest management decisions.

The models should be used with the knowledge that they are an approximation of the risk and the location of the risk posed by *O. erosus*. Local knowledge should always be considered when using these products. The relatively large knowledge and data gaps prevent these products from being completely precise. Indeed, they were constructed with data that was not entirely collected for these purposes and based on expert interpretation of incomplete knowledge about the invasive pest and its susceptible hosts. In addition, information as to where imports are coming from and ending up is not available. The applicability of the products is also limited in the scope of time; they describe risk in the short-term, and these risks may change with even the passage of a few years as environmental and societal conditions change.

Summary

The *O. erosus* mapping project demonstrates the interdependence of R&D and FHP staff and provides a clear example of how the two groups can work both cooperatively and synergistically to:
1. conduct timely research;
2. attain needed population and biological information;
3. immediately implement research findings to develop tools; and
4. create tools that are useful to forest management personnel for taking appropriate actions in the field.

The project has also illustrated the benefit to government agencies of retaining some agility in directing resources toward developing problems. When the Washington Office FHP staff identified the need for tools to better understand the potential impact of *O. erosus*, the Pacific Southwest and Northern Research Stations were also identifying the need to improve the state of the science for *O. erosus* in North America. Research funding was in limited supply, so FHTET provided seed funds to the research cooperators to attain the needed population and biological information as well as to characterize the behavior of *O. erosus* in North America. In so doing, personnel with the appropriate support mechanisms and skill-sets, were brought together to develop needed tools.

The *O. erosus* Potential Susceptibility Map will be completed in early 2009, and the final products will be posted on the FHTET web site (USDA FS 2008), where the current and future status can be monitored by users. Final products will include three maps (the Potential Introduction, the Potential Establishment, and the Potential Susceptibility); a recommended survey sampling design for the U.S. (survey sample areas); links to the biological attributes of *O. erosus* (*via* the ExFor site, http://spfnic.fs.fed.us/exfor); links to key pieces of scientific literature (e.g., Lee et al. 2005, 2008, In press); a list of forest species (hosts) at risk; the methods used in developing the Susceptibility Potential Map; the metadata; and the membership of the steering committee.

For the first time in the history of forest insect investigations in the western U.S., invasive subcortical pests from other continents have established populations that threaten conifers. The appearance of *O. erosus* and *H. ligniperda* in the urban forests of California will likely impact our future management of urban and peri-urban pines in this state and beyond if the populations expand. The capacity of the new invaders to compete with native populations of bark beetles in pines will be a research question of considerable interest (Amezaga and Rodríguez 1998). This new period of invasion of western U.S. forests by exotic subcortical insects has presented and continues to present an opportunity for USDA Forest Service R&D and FHP to pool their talents and resources to address a problem of pressing national concern.

Acknowledgments

We appreciate critical reviews of an earlier version of the manuscript by Robert A. Haack and Therese M. Poland (both USDA Forest Service, Northern Research Station, East Lansing, MI), John E. Lundquist (USDA Forest Service, Pacific Northwest Research Station, Anchorage, AK), and Robert C. Venette (USDA Forest Service, Northern Research Station, St. Paul, MN). We also thank Jana C. Lee, Deguang Liu, Mary Louise Flint (all UC-Davis, Dept. of Entomology), Frank H. Koch (North Carolina State University), Sujin Kim and Thomas C. Harrington (Department of Plant Pathology, Iowa State University), Whitney Cranshaw and Ned A. Tisserat (Colorado State University, Dept. of Bioagricultural Sciences and Pest Management), Richard L. Penrose (California Department of Food and Agriculture, Sacramento), Tom W. Coleman (USDA Forest Service, Forest Health Protection, San Bernardino, CA), John W. Coulston and William D. Smith (USDA Forest Service), and R.C. Venette for permission to cite or discuss unpublished data and for assistance with research on the invasion biology of *O. erosus* and other invasive species. Much of the information discussed in this paper was synthesized from several recent publications from Seybold's laboratory (Lee et al. 2005, 2007, 2008, In press; Liu et al. 2007, 2008) and we wish to acknowledge the numerous contributions of our co-authors and cooperators to them. Funding sources for these publications and associated research efforts came from USDA Forest Service Pacific Southwest Research Station; a University of California IPM Exotic/Invasive Pests and Diseases grant #05XU039 to SJS, M. L. Flint, and J. C. Lee; USDA NRI CSREES postdoctoral grant #2006-35302-16611 to J. C. Lee; USDA FS FHTET project "Dispersal potential of the Mediterranean pine engraver, *Orthotomicus erosus*, in support of risk mapping," to SJS, R.C. Venette, and M.L. Flint; and USDA Forest Service Special Technology Development Program grant "Improved early detection for the Mediterranean pine engraver, *Orthotomicus erosus*, an invasive bark beetle (R4-2008-01)," to SJS, A.S. Munson, R.C. Venette, and M.L. Flint.

Literature Cited

Amezaga, I.; Rodríguez, M.Á. 1998. Resource partitioning of four sympatric bark beetles depending on swarming dates and tree species. Forest Ecology and Management. 109: 127–135.

Baylis, N.T.; de Ronde, C.; James, D.B. 1986. Observations of damage of a secondary nature following a wild fire at the Otterford State Forest. South African Forestry Journal. 137: 36–37.

Ben Jamaa, M.L.; Lieutier, F.; Yart, A.; Jerraya, A.; Khouja, M.L. 2007. The virulence of phytopathogenic fungi associated with the bark beetles *Tomicus piniperda* and *Orthotomicus erosus* in Tunisia. Forest Pathology. 37: 51–63.

Bevan, D. 1984. *Orthotomicus erosus* (Wollaston) in Usutu pine plantations, Swaziland. Forest Research Report Number 64. Usutu Pulp Company Limited. 34 p.

Bright, D.E., Jr.; Stark, R.W. 1973. The bark and ambrosia beetles of California, Coleoptera: Scolytidae and Platypodidae. Bulletin of the California Insect Survey, Vol 16. Berkeley, CA: University of California Press. 169 p.

Cathey, H.M. 1990. Plant hardiness zone Map. Miscellaneous Publication No. 1475. Washington, DC: US Department of Agriculture, U.S. National Arboretum, Agricultural Research Service. http://www.usna.usda.gov/Hardzone/ushzmap.html (12 January 2009).

Coleman, T.W.; Seybold, S.J. [In press]. Previously unrecorded damage to oak, *Quercus* spp., in southern California by the goldspotted oak borer, *Agrilus coxalis* Waterhouse (Coleoptera: Buprestidae). The Pan-Pacific Entomologist.

Coleman, T.W.; Seybold, S.J. 2008. New pest in California: The goldspotted oak borer, *Agrilus coxalis* Waterhouse. Pest Alert, R5-RP-22. Vallejo, CA: U.S. Department of Agriculture, Forest Service, Pacific Southwest Region, State and Private Forestry. 4 p.

Dodds, K.J.; Gilmore, D.W.; Seybold, S.J. 2004. Ecological risk assessments for insect species emerged from western larch imported to northern Minnesota. Department of Forest Resources Staff Paper Series Number 174. St. Paul, MN: University of Minnesota. 57 p.

Eglitis, A.E. 2000. Mediterranean pine engraver beetle. In: Pest risk assessment for importation of solid wood packing materials into the United States. Washington, DC: U.S. Department of Agriculture, Animal and Plant Health Inspection Service and Forest Service: 190–193.

Environmental Systems Research Institute (ESRI). 2008. ArcGIS® Model Builder® Spatial Analyst®. 380 New York Street, Redlands, CA.

Furniss, M.M.; Johnson, J.B. 2002. Field guide to the bark beetles of Idaho and adjacent regions. Moscow, ID: University of Idaho, Idaho Forest, Wildlife, and Range Experiment Station. 125 p.

Furniss, R.L.; Carolin, V.M. 1977. Western forest insects. Misc. Pub. 1339. Washington, DC: U.S. Department of Agriculture, Forest Service. 654 p.

Gandhi, K.J.K.; Gilmore, D.W.; Katovich, S.A.; Mattson, W.J.; Spence, J.R.; Seybold, S.J. 2007. Physical effects of weather events on the abundance and diversity of insects in North American forests. Environmental Reviews. 15: 113–152.

Goheen, D.J.; Hansen, E.M. 1993. Effects of pathogens and bark beetles on forests. In: Showalter, T.D.; Filip, G.M., eds. Beetle-Pathogen Interactions in Conifer Forests. London: Academic Press: 175–196.

Grulke, N.E.; Preisler, H.K.; Rose, C.; Kirsch, J.; Balduman, L. 2002. O_3 uptake and drought stress effects on carbon acquisition of ponderosa pine in natural stands. New Phytologist. 154: 621–631.

Haack, R.A. 2001. Intercepted Scolytidae (Coleoptera) at U.S. ports of entry: 1985–2000. Integrated Pest Management Reviews. 6: 253–282.

Haack, R.A. 2006. Exotic bark- and wood-boring Coleoptera in the United States: recent establishments and interceptions. Canadian Journal of Forest Research. 36: 269–288.

Halperin, J.; Mendel, Z; Golan, Y. 1983. On the damage caused by bark beetles to pine plantations: Preliminary report. La-Yaavan 33: 46.

Hicke, J.A.; Logan, J.A.; Powell, J.; Ojima, D.S. 2006. Changing temperatures influence suitability for modeled mountain pine beetle (*Dendroctonus ponderosae*) outbreaks in the western United States. Journal of Geophysical Research. 111: [Not paged] G02019, doi:10.1029/2005JG000101.

Jacobi, W.R.; Koski, R.D.; Harrington, T.C.; Witcosky, J.J. 2007. Association of *Ophiostoma novo-ulmi* with *Scolytus schevyrewi* (Scolytidae) in Colorado. Plant Disease. 91: 245–247.

Jiang, Y.-P.; Huang, Z.-Y.; Huang, X.-C. 1992. Studies on *Orthotomicus erosus*. Journal of Zhejiang Normal University (Natural Science) 15: 79–81 In Chinese.

Johnson, P.L.; Hayes, J.L.; Rinehart, R.E.; Sheppard, W.S.; Smith, S.E. 2008. Characterization of two non-native invasive bark beetles, *Scolytus schevyrewi* and *Scolytus multistriatus* (Coleoptera: Curculionidae: Scolytinae). The Canadian Entomologist. 140: 527–538.

Koch, F.; Smith, B.; Coulston, J. 2007. Mapping drought conditions using multi-year windows. 10 p. Unpublished report. On file with: U.S. Department of Agriculture, Forest Service, Forest Health Monitoring Program, Forest Sciences Laboratory, 3041 Cornwallis Road, Research Triangle Park, NC 27709.

Krist, F.J.; Sapio, F.J.; Tkacz, B. 2007. Mapping risk from forest insects and diseases. FHTET 2007–06. Washington, DC: US Department of Agriculture, Forest Service, Forest Health Protection Forest Health Technology Enterprise Team. 115 p.

Langor, D.W.; DeHass, L.J.; Footit, R.G. 2009. Diversity of non-native terrestrial arthropods on woody plants in Canada. Biological Invasions. 11: 5–19.

Lee, J.C.; Aguayo, I.; Aslin, R.; Durham, G.; Hamud, S.M.; Moltzan, B.D.; Munson, A.S.; Negrón, J.F.; Peterson, T.; Ragenovich, I.R.; Witcosky, J.J.; Seybold, S.J. [In press]. Co-occurrence of two invasive species: The banded and European elm bark beetles (Coleoptera: Scolytidae). Annals of the Entomological Society of America.

Lee, J.C.; Flint, M.L.; Seybold, S.J. 2008. Suitability of pines and other conifers as hosts for the invasive Mediterranean pine engraver (Coleoptera: Scolytidae) in North America. Journal of Economic Entomology. 101: 829–837.

Lee, J.C.; Haack, R.A.; Negrón, J.F.; Witcosky, J.J.; Seybold, S.J. 2007. Invasive bark beetles. Forest Insect & Disease Leaflet 176. U.S. Department of Agriculture, Forest Service. 12 p.

Lee, J.C.; Jiroš, P.; Liu, D.; Hamud, S.M.; Flint, M.L.; Seybold, S.J. [in prep]. Pheromone production and flight response to semiochemicals by the Mediterranean pine engraver (Coleoptera: Scolytidae), a recently invasive bark beetle in California. Journal of Chemical Ecology.

Lee, J.C.; Negrón, J.F.; McElwey, S.J.; Witcosky, J.J.; Seybold, S.J. 2006. Banded elm bark beetle - Scolytus schevyrewi. Pest Alert, R2-PR-01-06. Golden, CO: U.S. Department of Agriculture, Forest Service, Rocky Mountain Region, Forest Health Protection. 2 p.

Lee, J.C.; Smith, S.L.; Seybold, S.J. 2005. The Mediterranean pine engraver, Orthotomicus erosus. Pest Alert, R5-PR-016. Vallejo, CA: U.S. Department of Agriculture, Forest Service, State and Private Forestry, Pacific Southwest Region. 4 p.

Little, E.L., Jr. 1971. Atlas of United States trees, Volume 1, Conifers and important hardwoods. Miscellaneous Publication 1146. Washington, DC: U.S. Department of Agriculture. 400 p.

Liu, D.-G.; Bohne, M.J.; Lee, J.C.; Flint, M.L.; Penrose, R.L.; Seybold, S.J. 2007. New introduction in California: The redhaired pine bark beetle, Hylurgus ligniperda Fabricius. Pest Alert, R5-PR-07. Vallejo, CA: US Department of Agriculture Forest Service, State and Private Forestry, Pacific Southwest Region. 3 p. http://www.fs.fed.us/r5/spf/publications/pestalerts/HylurgusPertAlert.pdf.

Liu, D.-G.; Flint, M.L.; Seybold, S.J. 2008. A secondary sexual character in the redhaired pine bark beetle, Hylurgus ligniperda Fabricius (Coleoptera: Scolytidae). The Pan-Pacific Entomologist. 84: 26–28.

Mattson, W.J.; Lawrence, R.K.; Haack, R.A.; Herms, D.A.; Charles, P.-J. 1988. Defensive strategies of woody plants against different insect-feeding guilds in relation to plant ecological strategies and intimacy of association with insects. In: Mattson, W.J.; Levieux, J.; Bernard-Dagan, C., eds. Mechanisms of woody plant defenses against insects: search for patterns. New York: Springer Verlag: 3–38.

Mattson, W.J.; Niemela, P.; Millers, I.; Inguanzo, Y. 1992. Immigrant phytophagous insects on woody plants in the United States and Canada: an annotated list. Gen. Tech. Rep. NC-GTR-169. U.S. Department of Agriculture, Forest Service, North Central Forest Experiment Station. 27 p.

Mendel, Z. 1983. Seasonal history of *Orthotomicus erosus* (Coleoptera: Scolytidae) in Israel. Phytoparasitica. 11: 13–24.

Mendel, Z.; Halperin, J. 1982. The biology and behavior of *Orthotomicus erosus* in Israel. Phytoparasitica. 10: 169–181.

Mendel, Z.; Boneh, O.; Riov, J. 1992. Some foundations for the application of aggregation pheromone to control pine bark beetles in Israel. Journal of Applied Entomology. 114: 217–227.

Mendel, Z.; Boneh, O.; Shenhar, Y.; Riov, J. 1991. Diurnal flight patterns of *Orthotomicus erosus* and *Pityogenes calcaratus* in Israel. Phytoparasitica. 19: 23–31.

Moser, J.C.; Fitzgibbon, B.A.; Klepzig, K.D. 2005. The Mexican pine beetle, *Dendroctonus mexicanus*: first record in the United States and co-occurrence with the southern pine beetle—*Dendroctonus frontalis* (Coleoptera: Scolytidae or Curculionidae: Scolytinae). Entomological News. 116: 235–243.

Mouton, M.; Wingfield, M. J.; Van Wyk, P. S.; Van Wyk, P. W. J. 1994. *Graphium pseudormiticum* sp. nov.: A new hyphomycete with unusual conidiogenesis. Mycological Research. 98: 1272–1276.

NAPPFAST. 2008. North Carolina State University-Animal Plant Health Inspection Service: Weather based plant pest prediction system. http://www.nappfast.org/ (12 January 2009).

Negrón, J.F.; Witcosky, J.J.; Cain, R.J.; LaBonte, J.R.; Duerr, D.A., II; McElwey, S.J.; Lee, J.C.; Seybold, S.J. 2005. The banded elm bark beetle: a new threat to elms in North America. American Entomologist. 51: 84–94.

Paine, T.D.; Raffa, K.F.; Harrington, T.C. 1997. Interactions among scolytid bark beetles, their associated fungi, and host conifers. Annual Review of Entomology. 42: 179–206.

Parker, T.J.; Clancy, K.M.; Mathiasen, R.L. 2006. Interactions among fire, insects and pathogens in coniferous forests of the interior western United States and Canada. Agricultural and Forest Entomology. 8: 167–189.

Romón, P., Iturrondobeitia, J.C.; Gibson, K.; Lindgren, B.S.; Goldarazena, A. 2007. Quantitative association of bark beetles with pitch canker fungus and effects of verbenone on their semiochemical communication in Monterey pine forests in northern Spain. Environmental Entomology. 36: 743–750.

Scriven, G.T.; Reeves, E.L.; Luck, R.F. 1986. Beetle from Australia threatens eucalyptus. California Agriculture. 40(7/8): 4–6.

Seybold, S.J.; Huber, D.P.W.; Lee, J.C.; Graves, A.D.; Bohlmann, J. 2006a. Pine monoterpenes and pine bark beetles: a marriage of convenience for defense and chemical communication. Phytochemistry Reviews. 5: 143–178.

Seybold, S.J.; Lee, J.C.; Luxová, A.; Hamud, S.M.; Jiroš, P.; Penrose, R.L. 2006b. Chemical ecology of bark beetles in California's urban forests. In: Hoddle, M.S.; Johnson, M.W., eds. Proceedings of the 5th Annual Meeting of the California Conference on Biological Control, Riverside, California, July 27–28, 2006. [Publisher unknown]: 87–94.

Solomon, J.D. 1995. Guide to insect borers in North American broadleaf trees and shrubs. Agric. Handbook 706. Washington, DC: U.S. Department of Agriculture, Forest Service. 735 p.

U.S. Department of Agriculture, Forest Service. 2008. Forest Health Technology Enterprise Team. Invasive species information program; invasive species risk maps. http://www.fs.fed.us/foresthealth/technology/invasive_species.shtml (12 January 2009).

Venette, R.C.; Walter, A.J.; Seybold, S.J. 2009. Comparing risks from native and exotic bark beetles to the health of Great Lakes forests. In: Proceedings Society of American Foresters 2008 Annual Meeting. Reno, NV. CD-ROM. Society of American Foresters, Bethesda, MD.

Wood, S.L. 1982. The bark and ambrosia beetles of North and Central America (Coleoptera: Scolytidae), a taxonomic monograph. Provo, UT: Great Basin Naturalist Memoirs 6. 1359 p.

Wood, S.L. 2007. Bark and ambrosia beetles of South America (Coleoptera, Scolytidae). Brigham Young University, Provo, UT: M.L. Bean Life Science Museum. 900 p.

Yin, H.-F.; Huang, F.-S.; Li, Z.-L. 1984. Economic Insect Fauna of China, Fasicle 29, Coleoptera: Scolytidae. Beijing, China: Science Press: 138–139. In Chinese.

Zhou, X.-D.; Burgess, T.; de Beer, Z.W.; Wingfield, B.D.; Wingfield, M.J. 2002. Development of polymorphic microsatellite markers for the tree pathogen and sapstain agent, *Ophiostoma ips*. Molecular Ecology Notes. 2: 309–312.

Zhou, X.-D.; de Beer, Z.W.; Wingfield, B.D.; Wingfield, M.J. 2001. Ophiostomatoid fungi associated with three pine-infesting bark beetles in South Africa. Sydowia. 53: 290–300.

Zwolinski, J.B.; Swart, W.J.; Wingfield, M.J. 1995. Association of *Sphaeropsis sapinea* with insect infestation following hail damage of *Pinus radiata*. Forest Ecology and Management. 72: 293–298.

Pacific Northwest Research Station

Web site	http://www.fs.fed.us/pnw
Telephone	(503) 808-2592
Publication requests	(503) 808-2138
FAX	(503) 808-2130
E-mail	pnw_pnwpubs@fs.fed.us
Mailing address	Publications Distribution
	Pacific Northwest Research Station
	P.O. Box 3890
	Portland, OR 97208-3890

www.ingramcontent.com/pod-product-compliance
Lightning Source LLC
Chambersburg PA
CBHW080640180526
45168CB00008B/3239